Interactions of Man
and His Environment

Interactions of Man and His Environment

Proceedings of the Northwestern University Conference
held January 28-29, 1965

Edited by

Burgess H. Jennings
Associate Dean of Engineering
Northwestern University
Evanston, Illinois

and

John E. Murphy
Manager, Research Services
Borg-Warner Corporation
Roy C. Ingersoll Research Center
Des Plaines, Illinois

℗ PLENUM PRESS · NEW YORK · 1966

First Printing – October 1966
Second Printing – May 1968

Library of Congress Catalog Card Number 66-16371

© 1966 Northwestern University
Evanston, Illinois

Plenum Press
A Division of Plenum Publishing Corporation
227 West 17 Street, New York, N. Y. 10011

Printed in the United States of America

Preface

This symposium was a joint effort, co-sponsored by Northwestern University and Borg-Warner Corporation and included representatives of the latter-day estates of university, industry, and government. All of these groups were deeply interested in participating, since the problems of man and his environment, especially his urban environment, are ones which can be solved only through a cooperative effort.

The symposium proceedings present a review of the present position of man and his environment and outline the physical and social-science efforts being made to solve the problems posed by man's rapidly changing environment. The format of this book is such that Dr. Luther L. Terry's speech, "Environmental Health: Everybody's Business," serves as an introduction to the many-faceted discussions on the interactions of man and his environment. Mr. P. B. Gordon's speech "Man and His Environment — Where Are We? Where Are We Going?" aptly summarizes this symposium and serves this purpose in this publication.

Every effort was exerted to make this symposium a general discussion meeting where contributions from both the formal speakers and the audience were received. The records of two panel discussions, including comments from the floor and replies by the speakers, are included in this volume.

This publication is the result of the coordinated efforts of many contributors: the speakers, Northwestern University, Borg-Warner Corporation, the Symposium Planning Committee consisting of D. W. Collier, G. F. Hanna, E. R. Hermann, J. E. Jacobs, B. H. Jennings (Chairman), R. L. Kuehner, J. E. Quon, myself, and especially my secretary, Catherine O'Hare, who spent many hours compiling and retyping the various papers. I also wish to thank my wife, Lorraine, for her encouragement and understanding.

Borg-Warner Corporation
Roy C. Ingersoll Research Center
Des Plaines, Illinois   John E. Murphy

Foreword

Since the days when prehistoric man huddled in his cave, irritated by the smoke from his flickering fire, environmental control has challenged man. Even as the caveman had to endure smoke while he worked to improve his environment, so modern man has continued to complicate his situation by adding pollutants to the atmosphere and to water supplies. Nature with its largesse has given us enormous amounts of air and water, and yet the availability of uncontaminated air and water for growing population needs has become a real problem in many parts of the world. More research and design planning, with proper conservation practice, will be required if successful solutions are to be envisaged. Thus, it is indeed fitting and appropriate for engineers, scientists, and doctors to meet at this conference and take stock of the measures needed to make the earth a better place for man to live.

Research has not effectively kept up with the creation of controlled countermeasures to offset the effects of pollution. This is true in terms of industrial wastes where smoke, fumes, and dusts taint the air and where increasing amounts of pollutants pour into the streams and damage watersheds. However, industry is not responsible for the pollution created by individuals with improperly controlled heating systems, for the burning of trash and leaves, and for poorly-adjusted automobile exhausts, and all of these contribute their share of fumes to the air. Moreover, many communities work with inadequate sewage treatment facilities. In some areas of the country, the problems have reached crisis proportions. Fully effective countermeasures have not been developed nor are they in sight unless effective research is carried out.

The toxic effects of pollutants in gross quantities are easily recognized, and the very excess of them forces countermeasures into play. The insidious effects of air pollution are hidden when the toxic matter reaches the human in such small amounts that only over a period of time does real danger come to light. The medical profession is working diligently to discover the cumulative effects of toxic components of the atmosphere, and in many cases maximum allowable concentrations are known, so that countermeasures can be put into effect to keep toxicity below such levels.

The picture of man and his environment is not altogether a bleak one, since great strides have been made by the air-conditioning industry so that the indoor environment at least can be controlled within close limits to provide safe and comfortable atmospheres. The time when air conditioning was considered only for special applications or for the priviledged few has long since passed and air conditioning is now a reality for almost every type of indoor occupancy. Even here, however, information as to true optimum conditions for occupancy still presents some uncertainties, and continuing research needs to be carried out. Even more-challenging areas of endeavor involve man in space and man under the sea. Fortunately these little-known areas involve limited numbers of people, but even here research needs to be continued if man hopes to conquer space and the bounty of the sea. Here let us call attention to the unsolved problem of man's environment and point our efforts to the solution of them.

Northwestern University   Burgess H. Jennings
Evanston, Illinois

Foreword

One of the prime objectives of this conference has been to bring together some of the foremost experts on the problems of environment and promote an interchange of their ideas on the subject.

Borg-Warner has been fortunate in being able to have helped sponsor this symposium and our reasons are not mainly self-serving. We are concerned that, in the midst of our population explosion, man's uncontrolled interactions with his environment are rapidly reaching the crisis state. It has been our hope that the interchange of ideas resulting at this conference would uncover new insights that would point out directions in which solutions might be found to our ever growing environmental problems.

Now a word about why Borg-Warner wanted to play an important role in bringing about such a conference. We have had a lively interest for some time in a variety of environmental projects, including activity in air conditioning, heating, refrigeration, air purification, and hyperbaric oxygen medicine.

The controlled environmental chamber of the Borg-Warner Science Hall is a unique demonstration of what can be done in the direction of total environmental control. Inside the environment chamber, audiences feel the heat of the summer sun and the chill of winter snows, as temperature, humidity, odors, and colors inside the environment chamber change to match the action on the screen. They smell the odor of the musty cave that was man's first shelter, the field of flowers, a wet dog, and a girl's perfume. Colors inside the chamber change from hot reds to cool blues when the temperature drops. These quick changes in atmosphere serve to dramatize recent advances in air sanitation and purification.

Thus, this working example of total environment control emphasizes what has been our basic approach. The problem of environmental control is one which should be looked at on an integrated basis, one with the ultimate goal being the total control of environment to secure the optimum in health and comfort conditions.

Borg-Warner Corporation　　　　　　　　　　　　　　　　　D. W. Collier
Chicago, Illinois

Contents

Environmental Health: Everybody's Business

Luther L. Terry, M.D.
Surgeon General, Public Health Service
U.S. Department of Health, Education, and Welfare
Washington, D.C.

In the field of environmental health, we are dealing, quite literally, with man's future in the society he has created for himself. And those of us who are working in that field are writing prescriptions for survival in a world grown incredibly intricate and complex.

That is why I am happy to participate in this conference of scientists, teachers, administrators, and health leaders on the inter- actions of man and his environment. To me, there are few more important themes for us to explore in considering the direction of human development.

President Johnson summed up both our goal and our challenge when he told the Country in his State of the Union Message that "an educated and healthy people require surroundings in harmony with their hopes."

In the Great Society, he said, "we want to grow and build and create, but we want progress to be the servant and not the master of man."

In a very profound sense, this is the essence of environmental health in today's world, to help the society of man reap the benefits of modern science and at the same time to protect ourselves against its possible hazards.

For those of us engaged in maintaining and furthering environmental health, harmonious surroundings mean clean air, water, food, and neighborhoods. They mean control of natural hazards, and those created by man himself, to assure human well-being. And they mean a healthy and significant life in our homes, our places of work, and in the way we use our leisure time.

Considered thus broadly, and I believe we must avoid limited and compartmentalized thinking, a healthful environment becomes a

1

basic social objective. Environmental health is a positive concept, designed not only to protect but to promote increasing levels of well-being.

My subject today is "Environmental Health: Everybody's Business." Since, at the most elementary level, we equate business with buying and selling, I am satisfied to start our discussion with the thought that environmental health *is* purchasable.

We can create clean and healthful communities; we can protect our population from food poisoning; we can recast jobs and places of employment so that work will prolong rather than shorten life. Our technical proficiency and scientific knowledge are equal to all these tasks.

I am less sure about our interest and our willingness, as individuals or as a society, to summon up the energy necessary and to pay the large bill required for such accomplishments.

Also, I am disturbed by the need, in dealing with the environment, to turn one outstanding liability into an asset. That difficulty was memorably stated by Izaac Walton who remarked: "I remember what a wise friend of mine did usually say, 'That which is everybody's business is nobody's business.' "

The environment is everybody's business, and this leads to a great number of difficulties. It is the business—the working life, if you will—of a variety of specialists: engineers, architects, radiologists, meteorologists, aquatic biologists, electron microscopists, toxicologists, systems analysts, planners, and many others.

It is the business of numerous agencies within and outside of government and at every level of government. Nearly every agency of government, for example, has some responsibility for studying the environment or for controlling one of its resources. Moreover, the community response to environmental problems in our coalescing urban areas depends on satisfactory adjustments within an intricate administrative system, involving hundreds of local jurisdictions.

And it has become the business of the American people, who no longer are willing to tolerate unwise, uneconomic, and unhealthful use of the environment. This public interest is often manifested through voluntary organizations which call for quick solutions to many complex environmental problems.

We must find a way to harness this interest and this energy and, at the same time, pursue an orderly and statesmanlike course. This requires a high degree of cooperation and coordination.

Coordination is one of the most difficult arts of administration. Yet, coordination is essential in an operation which stretches across many boundaries, geographic and administrative and philosophical. Since coordination is never perfect, the scope of interest in and involvement with the environment poses one of our first major problems.

The environment is everybody's business also in the sense that we are all, in some measure, responsible for fouling the air, water, and soil around us. Exhaust from our automobiles, detergent from our kitchen sinks, and waste products from our business centers, agricultural areas, and industries are poured into the skies above and into the rivers nearby. Wastes from individual homes, factories, and farms are multiplied as our population expands in numbers and is concentrated into the great urban areas of the Nation.

I am sure we would all agree, therefore, that environmental health is urgent business. There are two reasons for this urgency, one of which might be termed "external" and the other "internal."

The external reason for urgency involves the pace of scientific and technological development and the increasing public demand for action.

We can get some idea of the nature and magnitude of the problems simply by listing some of the dominant trends in American society today: the growth of our population; the ever-increasing diversity of our technology; the development of new industries; the increasing use of nuclear power; the magnitude of the gross national product; the introduction of new chemicals into our food, water, air, and consumer products; the growth of huge metropolitan areas; altered means of communication and transportation; and increased leisure time and the demand for recreational facilities.

Moreover, the problems are multiplying with dazzling speed. Many U.S. cities, for example, experienced their first smog episodes within the past five to ten years. My home town of Washington, probably the least industrialized U.S. city of its size, had its first recorded instance of Los Angeles-type smog in June 1960; there have been several since.

In any field we examine, the story is the same. There is now six times as much pollution in our rivers, streams, and lakes as 60 years ago, and the amount is still increasing. Every year, more than 500 new chemicals and chemical compounds are introduced into industry, along with countless operational innovations.

Our challenge is no less urgent when we consider the oldest health problems of the environment: water supply, waste disposal, and general sanitation. In the present period, they have taken on such vast dimensions and have become involved with so many economic and political issues as to be classed as *new* city health problems.

For example, the collection and disposal of solid wastes will be a continuing problem as long as some three-fourths of all communities with over 1000 population use open dumps or other poor means of disposal. These antiquated methods contribute to pollution, the propagation of disease vectors, odors, and ugly conglomerations of garbage and junk. It is no wonder that the people are insisting on remedial action.

The public is also increasingly restive about paying the price of pollution, the price in dollars, in threats to their health, in blighted communities. The price is high; and it will grow higher unless we make the conscious commitment now to reverse the tide that threatens to engulf us.

The second, or "internal," reason for urgency is the state of our knowledge. We know a good deal about the environment, and our knowledge is growing every day. Most important, we know enough to be aware of the extent of our ignorance, and that is the real beginning of knowledge.

Several years ago a committee of distinguished scientists reported that this Nation was at least ten years behind schedule in its research effort on environmental contamination. Today, with the situation aggravating instead of diminishing, we still have little exact knowledge of what takes place within the human body when it inhales, ingests, or comes into physical contact with toxic substances in small quantities over a long period of time. Possible genetic effects of long-term exposure to potentially harmful substances can only be guessed at. Scientific protection against radiation is at the beginning stage of development. The effects of such physical forces as heat, cold, and noise are little known.

Throughout the centuries, man has shown great ability to adjust to varying environments. But there may be a limit to his ability to adapt, particularly in the face of the drastic and far-reaching changes of today's world.

Should we not be energetically, even frantically, at work to safeguard our environment and to protect our health? Of course we should be, yet we can identify some reasons why we are not moving as quickly as necessary.

One reason lies in the need to take corrective rather than preventive measures. It is very difficult to stop or remove pollution once it has a good start. At a recent air pollution control conference, one of the speakers put it this way: "Once a process has become embedded in a vast economic or political commitment, it may be nearly impossible to alter."

We become used to doing things a certain way. The costs of changing, or retooling, of reorganizing are high. Frequently these costs fall unequally, and the part of society which may have to shoulder the major portion of the cost tends to resist.

Another reason is that we become accustomed to what Mark Twain called "all the modern inconveniences." Once we become used to a condition, it is difficult to see it clearly as a danger.

Whose function is it, at this point, to bring the efficiency and energy to bear upon managing everybody's business, the environment? As in any smooth-running establishment affecting the well-being of the total population, we all have our roles to play.

Responsibility for national policy, of course, lies at the highest levels of government. President Johnson has said: "The Great Society which we mean to build in America must be a healthy society. I pledge my wholehearted energies to make it that way."

And he has repeatedly stressed his intention of creating a wholesome environment as one of the indispensable steps toward better health in the Great Society. In his Health Message to the Congress early this year—the first of his special messages, by the way, a fact indicative of the high priority he gives to health matters— and in his special message on National Beauty, he devoted considerable attention to the pollution of our environment.

The President's budget which was submitted to the Congress included an increase of $17 million for environmental health activities in the Public Health Service. With this increase, we hope to strengthen several of our programs, and to initiate several important new activities.

For example, we are currently conducting studies on the effects of pesticides on human health. We hope to expand the community studies already under way and to accelerate our research efforts in the analysis of the health effects of the continued and long-range use of pesticides.

We are planning a concentrated effort to combat botulism poisoning by intensifying our research and development of public health methods to identify and control poisoned foods. I am sure all of you remember the consternation which followed the cases of botulism you experienced here in the Midwest about a year ago. We must not permit such episodes again.

We hope to expand our attack on a major problem of our cities— solid waste disposal. For example, we plan to test and demonstrate a new process by which two substances, solid wastes and sewage sludge, both of which are useless and pose disposal problems, can be converted into something that will be useful as a soil conditioner. This project will be conducted in cooperation with the Tennessee Valley Authority. We hope that it will offer a 20th Century solution to an age-old, expensive, and increasingly serious problem.

In the fields of air and water pollution, we plan to increase our support of community programs, research, training, and technical services. There will be an increase in research activities in the determination of the health effects of sulfur compounds in the air, the removal of sulfur from fuels, and the development of air pollution control devices.   We also plan to expand the cooperative project initiated in 1965 to control water pollution from acid mine drainage.

Most important of all, in my opinion, we are taking steps to provide a focal point for far-ranging and coordinated research, training, and control programs in the environmental health sciences. I have just listed some components of our environmental health program, but I want to emphasize that we regard them as separate facets of the same problem.   The same chemicals, for example, may impinge on the individual in community air, in milk, foods, water, and in his occupational environment, not once, but repeatedly.

Obviously, different techniques may be required to cope with the threats present in different sectors of the environment. It may be necessary to manipulate the environment for broader purposes than the protection of the population against specific diseases, as is the case in water pollution control. But the man—environment relationship is "one and indivisible."

In the business of environmental health, therefore, we need to look beneath and beyond the immediate operating responsibilities of official agencies at all levels, as well as those of industries and scientific institutions.   And we need to keep man at the center of our considerations.

The Department of Health, Education, and Welfare recently began the final stage of planning a major center for the study of man and his environment. Called the National Center for Environmental Health Sciences, it will bring together a group of scientists and administrators to establish a national leadership effort in this field. The North Carolina Research Triangle, a hub of academic, scientific, and industrial activities, has been selected as the site of the Center, but it is also anticipated that additional environmental health facilities will be established in other locations in the United States.

One of the Center's primary functions will be to develop and maintain a national overview of the needs in environmental health research and to help fill in the gaps, either in its own laboratories or by contracts and grants to universities and other nongovernment institutions.   It will appraise and analyze our national requirements in the environmental health sciences and our directions for the future. To accomplish this task, the Center will require a wide range of competencies in many scientific fields.

We have high hopes for the Center, of course, as the nucleus for a major new thrust against the problems of the modern environment. But the problems are national in scope and call for a national response. Research workers in universities and laboratories throughout the country will continue to be a vital part of the total effort.

All of us have a share in developing techniques and resources that will contribute to a healthier environment. And all of us must recognize change, anticipate change, and work toward change.

The greatest change, one that is prerequisite to all others, is a change in our thinking and attitudes. John Erskine said: "The body travels more easily than the mind, and until we have limbered up our imagination we continue to think as though we had stayed home. We have not really budged a step until we take up residence in someone else's point of view."

Business as usual when the business at hand is the environment must yield to new philosophies and approaches.

Those of you who are university faculty members, research workers, engineers, scientists, and business men have a responsibility beyond that of working in your classroom, laboratory, or office. At this point in history, that responsibility is to contribute to public understanding of what is happening in our environment, and of what can and should be done in plotting our future course.

You are privileged to be engaged in a conference which, by its very content, calls attention to the importance to man of his environment. I hope you will convey the message to others as often and as clearly as possible.

This conference enables experts from different specialties to exchange views and to plan together. We need to provide such opportunities from time to time. We need to pause in our research and in our action programs to provide for the kind of thinking that will result in the application of knowledge from one field of activity to another.

We need to incorporate as part of our technological process an evaluation of the effects of new procedures or products on society as a whole. This moment of hesitation, this slight pause before a leap into action, may be the dose of preventive medicine that will give to millions good health in place of disability and death. It is the application of the philosophy of preventive medicine to society.

Protection of the environment is everybody's business—a most noble enterprise in which our profits are a healthier world for ourselves and for all the generations to follow.

Man's Relationship to His Thermal Environment

Douglas H. K. Lee, M.D.

Division of Occupational Health
Public Health Service
U.S. Department of Health, Education, and Welfare
Cincinnati, Ohio

The quantitative expression of man's reaction to his thermal environment poses a complex problem. There are three sets of variables, each containing several items or subsets which must be considered; for example:

- Environmental—temperature, humidity, air movement, radiant heat, clothing insulation, and contiguity;
- Individual—age, sex, body build, disease, hydration, level of activity, acclimatization, and individual variability;
- Evaluation (criteria for assessment of effect)—comfort-discomfort, sensation of distress, functional failure, pathological developments, aggravation of previous defects, susceptibility to other stresses, water requirements.

The problem of handling such a multiplicity of variables in meaningful fashion can be logically dealt with in five steps:

1. From heat transfer equations devise expression of inter-relationship between appropriate variables (activity level, temperature, humidity, radiant heat, air movement, clothing insulation), and their net significance for man.

Using Burton's [1, 2] equations and the concept of relative strain introduced by Belding and Hatch [3], Lee and Henschel [4] developed the following approximate relationships:

$$RS = \frac{M(I_{cw} + I_a) + 5.55(t_a - 35) + RI_a}{7.5(44 - p_a)}$$

9

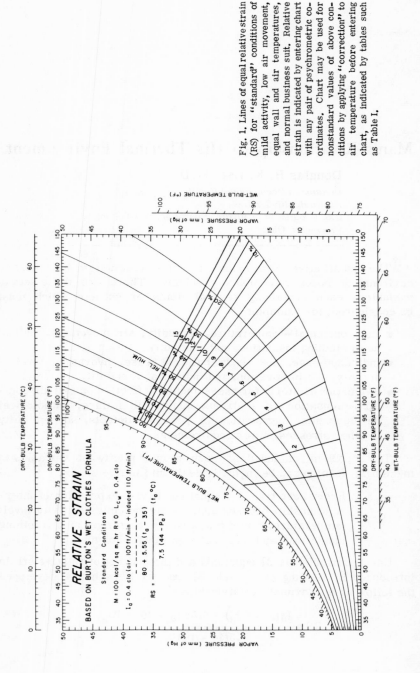

Fig. 1. Lines of equal relative strain (RS) for "standard" conditions of mild activity, low air movement, equal wall and air temperatures, and normal business suit. Relative strain is indicated by entering chart with any pair of psychrometric coordinates. Chart may be used for nonstandard values of above conditions by applying "correction" to air temperature before entering chart, as indicated by tables such as Table I.

where RS is relative strain (dimensionless); M is metabolic rate in kcal/m²-hr; R is radiant heat gain in kcal/m²-hr; t_a is air temperature in °C; I_a is insulation of air (inversely proportional to square root of air movement) in clo units; I_{cw} is insulation of wet clothes in clo units; and p_a is vapor pressure of air in mm Hg.

2. By postulating convenient "standard" values for metabolic rate (M), air movement (I_a), clothing insulation (I_{cw}), and radiant heat (R), reduce expression to effect of two independent variables (temperature and vapor pressure) on the dependent variable (relative strain), and draw lines of equal strain on a psychrometric chart.

Figure 1 gives such a chart for " standard" values of M = walking at 2 mi/hr; air movement 100 ft/min; wall temperature = air temperature; clothing = business suit.

3. From the expression for RS (relative strain), calculate the changes in air temperature which would produce the same changes in the value of RS as deviations from the "standard" values assigned to the variables of metabolic rate, air movement, radiant heat, and clothing insulation, and prepare tables to show "corrections" to actual air temperature which, if made before entering the chart, would compensate for such deviations.

One such table, for different values of activity and air movement, for use with Figure 1, is given as Table I.

Table I. Correction to Dry-Bulb Temperature (t_a) for Nonstandard Activity and Air Movement

Activity		Ambient air movement								
Type	M	25	50	75	100	200	300	400	500	ft/min
	kcal/m²-hr	8	15	23	30	61	92	122	153	m/min
Sleeping	40	-3	-3½	-4½	-5	-6	-6½	-7	-7	
Lying awake	50	-2	-2½	-3½	-4½	-5½	-6½	-7	-7	
Desk work	60	-3½	-4	-4½	-5	-5½	-6½	-7	-7	
Standing still	65	-1	-1½	-2	-2½	-3½	-4½	-5	-5	
Walking 2 mi/hr	100	0	0	-½	-1	-1½	-2	-2½	-2½	
Walking 3 mi/hr	140	+3½	+3	+2½	+2	+1½	+1	+½	-½	
Walking 4 mi/hr	190	+6	+5½	+5	+5	+4½	+4	+3½	+3	

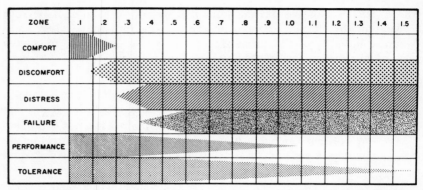

Figure 2. Interpretation of relative strain values for "standard" person (healthy, white male, about 25 years of age, accustomed to a median U.S. culture, with no particular acclimatization to heat). Zones are named for the relative strain values of their lower boundaries. Thickness of bar indicates average degree of probability of response.

4. From the data available in the literature and elsewhere, determine the probable effect of successive degrees of relative strain upon a defined "standard" person, in terms of selected evaluative criteria, and express in graphical form.

Figure 2 expresses the probable effect of each degree of relative strain from 0.1 to 1.5 upon a healthy, white male, age 25, living in a median U. S. Culture, and unacclimatized [4].

5. Prepare similar graphic expressions of probable effects for nonstandard persons.

The limited amount of useful data contained in the literature makes this last step difficult. An evaluation chart for a man who

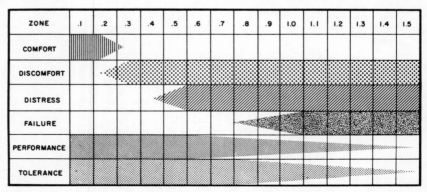

Figure 3. Interpretation of relative strain for acclimatized man. Note shift to right of response bars.

differs from the "standard" in being fully acclimatized is given as Figure 3. Others are given in a recent detailed paper [4].

The actual use of the scheme is comparatively simple:

a. From the appropriate table (such as that given as Table I) determine any "correction" to the actual air temperature needed to compensate for other than the "standard" values assigned to metabolic rate, air movement, radiant heat, or clothing.

b. With the air temperature and whatever measure of humidity is being used (wet-bulb temperature, relative humidity, or vapor pressure), enter the psychrometric chart (as in Figure 1). From the point so obtained, move horizontally to make the correction obtained from the table and read off the corresponding value of relative strain.

c. From the chart of effects appropriate to the persons under consideration (such as given in Figures 2 and 3) read off the probable effects indicated for that value of relative strain.

REFERENCES

1. Burton, A.C. "An analysis of the physiological effects of clothing in hot atmospheres," Rpt. of Aviation Med. Res. Assoc. Committee (Canada). C2754, SPC 186, 1944.
2. Burton, A.C., and Edholm, O.G. Man in a Cold Environment. Williams and Wilkins, Baltimore, 1955.
3. Belding, H.S., and Hatch, T.F. "Index for evaluating heat stress in terms of resulting physiological strains," Heat, Piping, Air Condit., Vol. 27, No. 8, pp. 129-136, 1955.
4. Lee, D.H.K., and Henschel, A. "Effects of physiological and clinical factors on response to heat," Ann. New York Acad. Sci. 134:743-749, 1965.

Man's Relationship to His Sensory Environment —Hearing and Vision

Max V. Mathews
Bell Telephone Laboratories
Murray Hill, New Jersey

Two of man's most important senses are hearing and vision. The great majority of the neurons in the brain deal with these senses. I will attempt to present some of their best-known characteristics, though in this limited space only a few of the most significant can be included. I will also describe some of their peculiarities that are particularly interesting to me. I will start with hearing.

A diagram of the ear is shown in Figure 1. A pressure wave in the air produces a mechanical motion of the eardrum, which is transmitted via a chain of three bones to the cochlea and produces a movement of the basilar membrane, thus exciting the hair cells which terminate the auditory nerve. The neural impulses so produced travel by a complex pathway through many parts of the lower brain and eventually produce measurable effects in the auditory cortex. The cochlea has been extensively studied, notably by von Békésy [1], who used the direct, if not simple, procedure of looking with a microscope at the vibrations of the basilar membrane through a small hole drilled in the cochlea. He was able to show that various frequencies of sound produce a maximum amplitude of vibration at various points on the basilar membrane and thus that the basilar membrane can act as a mechanical frequency-analyzing device. Curves of relative amplitude of vibration for six points on the basilar membrane as a function of frequency are shown in Figure 2. These data are consistent with the human ability to discriminate frequency. However, the rather broad maxima exhibited by these curves are insufficient by themselves to explain the very fine frequency discrimination which we achieve. In order to further study hearing, physiologists have inserted electrodes into the brain and

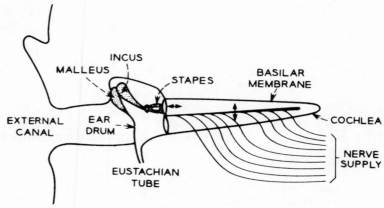

Fig. 1. Schematic diagram of the ear.

nerve cells and recorded the impulses which they found. Some typical data [2] from an electrode in the auditory nerve of a bat are shown in Figure 3. The solid curve shows the minimum sound intensity which will produce activity in one particular neuron as a function of the frequency of the sound. As can be seen, this particular neuron has its maximum sensitivity at a frequency of about 50 kc/sec. Parenthetically, I might add here that the bat hears at much higher frequencies than the human. The threshold curve here is very sharp, much sharper than the mechanical vibration curves shown in Figure 2. The question to be resolved is how the relatively broad tuning curves of mechanical motion can produce such a sharply peaked response from a particular neuron. One possible explanation put forward by Frishkopf is that adjacent sensory elements along the basilar membrane tend to inhibit each other, thus sharpening

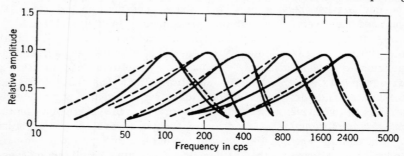

Fig. 2. Frequency response curves for six points on the cochlear partition. The solid curves are measured values (Békésy, 1943); the dashed curves are theoretical values calculated by Zwislocki (1948). (From Georg von Békésy and Walter A. Rosenblith, "The mechanical properties of the ear," handbook of Experience Psychology, S.S. Stevens [Ed.]. Wiley, New York, 1951, p. 1096.)

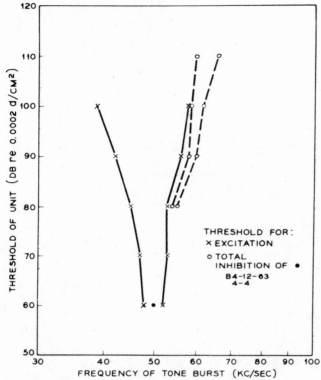

Fig. 3. Response of eighth nerve cell in a bat to tones applied to its ear. (Courtesy of L. S. Frishkopf.)

the effective tuning curve.   Frishkopf studied this inhibition by simultaneously applying two tones to the ear.   One tone was at 50 kc, the most sensitive frequency of the neuron; the other tone was at a slightly higher frequency.   The amplitude of the second tone sufficient to completely inhibit the firing of the neuron is shown by the dashed curve of Figure 3.   The region inside the dashed curve represents complete inhibition.   For frequencies only slightly away from 50 kc, a maximum amount of inhibition is obtained. The inhibition appears to occur in lateral connections among elements in the basilar membrane itself and can be used to explain the sharpening of the tuning curve.   In addition, similar inhibitory behavior has been found in the visual system and the touch sensations, so that it appears to be a general and important feature of sensory perception. I will refer to it again later.

The fundamental stimulus to hearing is, of course, a pressure wave, or pressure in the air, which changes as a function of time.

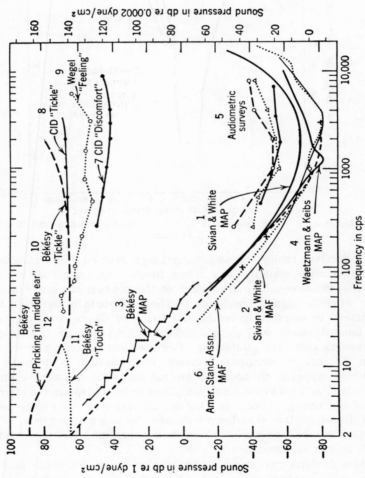

Fig. 4. Determinations of the threshold of audibility and the threshold of feeling. Curves 1 to 6 represent attempts to determine the absolute threshold of hearing at various frequencies. Curves 7 to 12 represent attempts to determine the upper boundary of the auditory realm, beyond which sounds are too intense for comfort and give rise to nonauditory sensations of tickle and pain. (From J.C.R. Licklider, "Basic correlates of the auditory stimulus," Handbook of Experimental Psychology, S.S. Stevens Ed, Wiley, New York, 1951, p. 955).

The frequency-analyzing characteristics of the cochlea which we have just seen make it reasonable to decompose the pressure wave into a sum of sinusoidal components. Figure 4, which gives a general view of the entire sinusoidal audibility space, is a composite diagram containing many peoples' data. The curves at the bottom of the graph indicate the threshold of audibility. Tones below these curves can, in general, not be heard; tones above these curves can be heard. The ear has its maximum sensitivity for frequencies of 1000 to 5000 cycles and rapidly-decreasing sensitivity at both higher and lower frequencies. The curves at the top of the graph indicate very approximately the threshold of pain. These curves do not vary greatly with frequency, and at very low frequencies, for example, the threshold of pain may occur at the same amplitude as the threshold of hearing. Sounds of amplitude sufficient to cause pain will, in general, damage the ear if they are sustained for any length of time. In fact, some sounds of considerably lower amplitudes, around 90 db SPL, can also produce damage if they are heard for a long time. At the most sensitive frequency of about 2000 cycles, the ratio of energy between a sound that can barely be heard and a sound just producing pain is 140 db or 10^{14}, which is an incredibly large number. Such a wide range of sensitivity is also characteristic of the visual system. It is also essential in our environment, since sounds both at threshold and at the threshold of pain frequently occur and are important to us.

Thresholds, either for audibility or pain, are a relatively simple concept. The loudness of a tone is a more complicated matter. Loudness is a subjective quality concerned with the perception of tones by a listener and should not be confused with the intensity of a sound which is a physical measurement. Two sorts of questions can be asked of a subject. First, what is the intensity of a tone which sounds exactly twice as loud as some other tone and, second, what is the intensity of a tone at one frequency which is just as loud as a tone at another frequency? If these questions can be consistently answered by a group of listeners, then a scale of loudness as a function of frequency and intensity can be constructed. Figure 5 shows one such scale in a three-dimensional view. The curve at the bottom represents zero loudness or the threshold of audibility. Successively higher curves show variation in loudness with frequency, again the maximum loudness for a given intensity being obtained at about 1000 cycles. However, the variation in these curves with frequency is not nearly as great as that of the threshold curve. Such curves of loudness and their elaborations are of great value for the musician who must know how loud the sound of various combinations of instruments will be, and also for the airplane designer who must

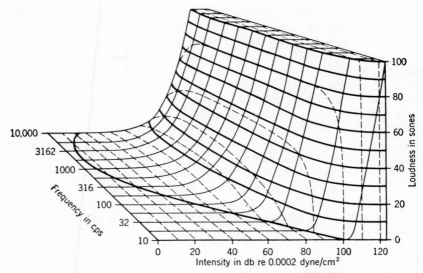

Fig. 5. Three-dimensional surface showing loudness as a function of intensity and frequency. Subjective loudness in sones is represented vertically above the intensity-frequency plane. The heavy curves coursing from front to rear in the diagram are equal-loudness contours for pure tones. (After Stevens and Davis, 1938.) (In J.C.R. Licklider, "Basic correlates of the auditory stimulus," Handbook of Experimental Psychology, S.S. Stevens [Ed.]. New York, 1951, p. 1002.)

predict the impact of his new jet on both its passengers and the people living around the airport.

So far we have been concerned with identical sound waves at both ears. Sound from a source in air travels about 1000 ft/sec and, consequently, may reach our two ears at slightly different times depending upon the angle of our head relative to the sound source. The maximum time difference is of the order of 500 μsec and, incredible as it seems, we have neural mechanisms for measuring time differences an order of magnitude less than this, as little as 10 μsec. Further, we use these time differences to determine the location of sound sources quite precisely. Figure 6 shows that we can locate a sound source of low frequency to an accuracy of perhaps 10° and sources of high frequency to somewhat lesser accuracy. Actually, these data have been superseded by later studies done by R. L. Hanson [3], which had a suprising result. Hanson found that, using very stable oscillators connected to excellent loudspeakers in a special anechoic chamber, localization was considerably poorer than that shown in Figure 6 for all frequencies. Somewhat by accident, he repeated the same experiment using a tape recorder rather than an oscillator to drive the loudspeakers. In this case, an immediate and dramatic improvement in localization ability occurred. Even the best tape recorders introduce a considerable amount of flutter or amplitude variation in the output sound and Hanson was able to show that we localize primarily not on the steady-state sound but, rather, on variations in amplitude or transients. This result has many interesting ramifications. Not only are we most strongly affected by changes in sounds, but we also give great priority in our attention to sounds which arrive first. For example, when two loudspeakers are set up, one to the left and one to the right, if the left speaker is turned on suddenly and then its intensity is gradually diminished and at the same time the right-hand speaker is turned on very gradually, a listener will continue to believe the sound originates from the left-hand speaker. Such a property can be effectively utilized to localize a soloist in a stereophonic loudspeaker system. However, this property contributed to some unexpected effects in Philharmonic Hall, recently built at Lincoln Center in New York City. As initially constructed, it was very difficult to hear the low-frequency instruments in the orchestra. Acoustic measurements showed only a slight attenuation in the total low-frequency energy being radiated from the orchestra to points in the audience. A more careful measurement, however, showed that the low-frequency energy

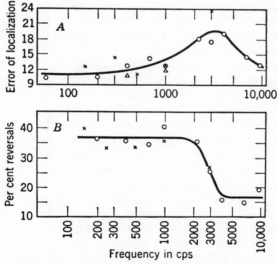

Fig. 6. Errors made in judging the direction of a sound source in the horizontal plane. Plot A shows the average of the errors, in degrees, for various tonal frequencies. The circles and the crosses are for two different series of observations with pure tone. The triangles are for impure tones. Plot B shows the percentage of confusions between front and rear quadrants. (After Stevens and Newman, 1936.) (In J.C.R. Licklider, "Basic correlates of the auditory stimulus," Handbook of Experimental Psychology, S.S. Stevens [Ed.]. Wiley, New York, 1951, p. 1027.)

tended to lag behind the high-frequency energy and that the initial sounds reaching the listener were deficient in low frequencies. The subjective effect of this moderately short lag of only a few tens of milliseconds was to almost obscure the base instruments. The physical reason for the delay was associated with a set of clouds, or acoustic reflectors, set between the ceiling and the floor. These were of such a size that the high-frequency energy reflected from them, whereas the low-frequency energy passed through the clouds and was reflected from the ceiling of the auditorium and, thus, arrived later. The difficulty was subsequently corrected by enlarging the clouds to form an almost completely connected false ceiling which reflects all frequencies of sound.

The peculiar phenomenon associated with localization of sounds is by no means exhausted. If you hear a sound while listening normally, you will localize the sound somewhere outside your head. On the other hand, if you listen to the same sound which is recorded by two microphones embedded in a dummy head, each microphone being connected to one earphone which you are wearing, the sound will be perceived as being localized somewhere between your two ears inside your head. The difference appears to be that your head is not kept stationary, and that you make use of the changes in sound produced by your head movements as a clue to externalize the sound source. Almost all sounds heard binaurally over earphones are thus localized internally. This factor is one of the significant things that should be considered in choosing between a hi-fi set with loudspeakers and one with earphones. Do you want the orchestra in your living room or in your head? Some people prefer one and some the other. I personally find my head is already a bit crowded.

Sounds presented over earphones are frequently used to make precise measurements of the hearing mechanism. Some data [4] about the precision of our ability to localize clicks presented binaurally are shown in Figure 7. Here clicks were presented to both ears at different intensities, and the subject could adjust the time of arrival of one of the clicks until the resulting sound image was in the center of his head. The accuracy of the adjustment is shown as a function of the difference in intensity between the two ears. The minimum error is $10\,\mu\mathrm{sec}$, which is very small indeed for human acoustic behavior. Such accuracy would be inconceivable if the brain depended on only one neural pulse. It must, instead, be taking a statistical average over the responses of many neurons.

Another result of significance which came from these data is a trading relation between time and intensity. In order to localize the sound, for example, on the right-hand side, it can either be presented earlier to the right ear, or it can be presented at a higher intensity.

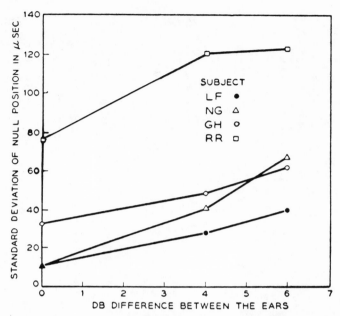

Fig. 7. The standard deviation of determining a null position of a binaurally present click as a function of the inter-aural intensity difference. (From Gerard G. Harris, "Binaural interactions of impulsive stimuli and pure tones," J. Acoust. Soc. Am., Vol. 32, p. 687, June 1960.)

Thus, intensity and time are equivalent in this sense. Whether this equivalence is caused by more rapid propagation of high-intensity pulses through the neural system or caused by some other feature, we do not know. However, there is no question but that we can take advantage of this ability. The adjustment for centering on a stereophonic system is simply one of intensity; one merely increases the volume of the right-hand speaker to move the sound image to the right.

I will mention one final effect of binaural hearing called the "cocktail party effect." This is the ability often noticed in very noisy places to concentrate on one particular voice and exclude the background noise. The effectiveness is easily demonstrated by comparing the difficulty in understanding a monaural recording of a conversation heard in a noisy background with understanding the actual conversation itself when you are in the environment. The effect is a complicated one; it involves not only directional hearing, but also head motion, external localization, and the ability to recognize a given voice, and probably the ability to fill in missing parts of sentences from the context of the rest of the utterance. However, it is an ability which we have developed to a very high degree and without which we would feel strongly handicapped.

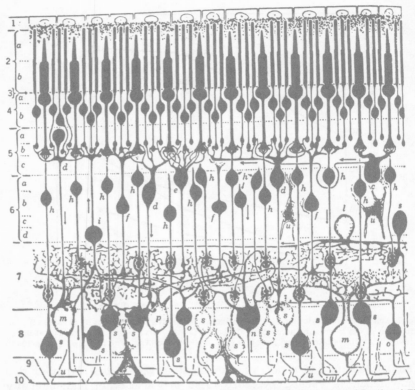

Fig. 8. The structure of the primate retina. (Polyak, 1941.) (In S. Howard Bartley, "The psychophysiology of vision," Handbook of Experimental Psychology, S. S. Stevens [Ed.]. Wiley, New York, 1951, p. 926.)

A question which we are asking in the telephone company is what degree of binaural communication must we provide for telephone conferences. Here it may be important to localize the talker at the distant conference table, which you cannot see, in order not to confuse various people and in order to clearly hear the conversation through whatever background noise exists. Whether stereophonic communication will be essential is as yet unanswered, but we have developed at least one effective technical means for presenting these sounds if it proves desirable.

I will now turn to the visual sense. Vision is undoubtedly our most important sense with regard to getting information from the external world. It is the sense which is most greatly missed when absent, and which apparently has the most complicated enervation and preempts more of the brain, if such a statement has meaning. Light enters the eye through a crystalline lens, which focuses an image on the light-sensitive retina at the back of the eyeball. Neural impulses from the retina proceed via the optic nerve, eventually arriving at the visual cortex. The retina itself is a very complicated structure, its neural anatomy being shown in simplified form in Figure 8. The figure shows a much-enlarged view of the very thin retina, with the crystalline lens below the section. Ten different layers of neurons are identified in the figure. The backmost layer or layer furthest from the crystalline lens contains the light-sensitive elements. These are of two general types, rods and cones, and light must go completely through the retina to reach them. I use this figure not in any attempt to explain the action of the retina, but to indicate its complexity and to make reasonable the conclusion that quite a bit of processing of the visual image is done in the retina itself and that the image transmitted to the optic cortex is far from a simple pictorial mapping of the picture projected onto the retina. The retina should be looked upon as an extension of the optic brain.

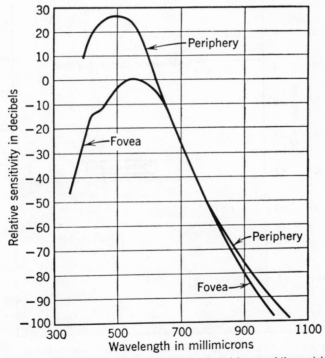

Fig. 9. The relative spectral sensitivity of the dark-adapted fovea and the peripheral retina. (From Griffin, Hubbard, and Wald, 1947.) (In S. Howard Bartley, "The psychophysiology of vision," Handbook of Experimental Psychology, S. S. Stevens [Ed.] . Wiley, New York, 1951, p. 935.

Many of the quantities which were appropriate to hearing can also be measured with respect to vision. For example, the sensitivity of the retina as a function of the frequency of the light wave is shown in Figure 9. Here the perception of frequency is color, the shorter wavelengths being perceived as violet light and the longer wavelengths as red light. Two curves are shown in the figure, one for the vision in a small area in the center of the eye, called the fovea, and one for vision elsewhere, called the periphery. The peripheral vision is primarily accomplished by rods, whereas the foveal vision is primarily a function of the cones, so that these two curves approxi-

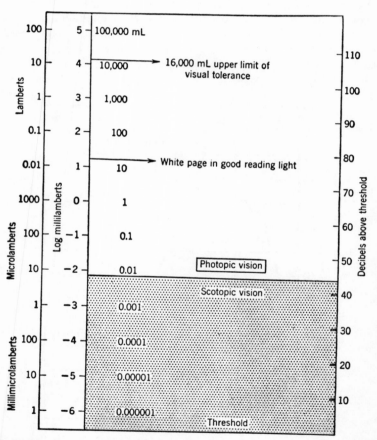

Fig. 10. Diagram indicating the extreme range of light intensities that the human eye confronts and the relation between the levels of cone and rod vision. The scale on the right shows that the range of visible intensities covers approximately 100 db, or a ratio of intensities equal to 10 billion to 1. (From S. Howard Bartley, "The psychophysiology of vision," Handbook of Experimental Psychology, S. S. Stevens [Ed.]. Wiley, New York, 1951, p. 945.)

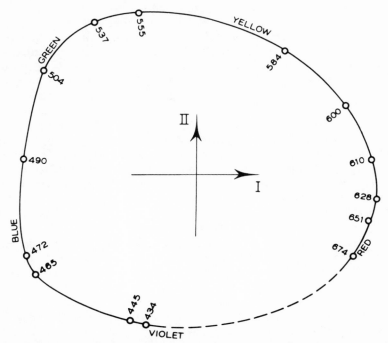

Fig. 11. Similarity space representing judgment of the similarity of pure colors. (From R.N. Shepard, "The analysis of proximities: Multidimensional scaling with an unknown distance function. II." Psychometrika, Vol. 27, p. 236, 1962.)

mately represent the difference in sensitivity between the cones and the rods. The rods have a maximum sensitivity at somewhat shorter wavelengths than the cones and, furthermore, they have a substantially greater overall sensitivity. Consequently, vision in very dim light is done primarily by the rod receptors. The cones, on the other hand, particularly those in the fovea, where the density of receptors is higher, have a much better spatial resolution than the rods, and detailed work such as reading is primarily a function of cone vision. The range of intensity to which the eye will accommodate is indicated in Figure 10 and spans a ratio of about 10^{10} in energy between the threshold for dark-adapted rod vision and the maximum illumination which can be tolerated without pain. This range, while slightly less than that of hearing, is still enormous. Furthermore, in contrast to what happens in a camera, in the eye this range of light intensity is in no significant way compensated for by changes in size of the iris.

Although there are many analogies between vision and hearing, there are also some basic differences. One of the most interesting of these is the relationship of time. The ear hears a pressure wave which varies as a function of time and in many ways is able to appreciate minute details of this time function. Although light may be thought of as an electromagnetic wave existing in time, there is no evidence that the eye can take advantage of any of the very fine-grained features of this wave, and the light can perhaps better be thought of as discrete events or photons impinging on the eye at random times. The efficiency of the eye in detecting these photons is excellent. Under the best conditions, only about 5×10^{-10} erg of energy is required to produce a detectable flash. This amounts to about 100 photons of energy. Assuming a normal amount of loss in the fluid in the eye and in the retina, we can estimate that probably only about 10 photons get through to the rods and cones; thus, they are indeed sensitive receptors. Someone has calculated that the energy represented by a pea falling one inch is sufficient to produce a flash visible to all the men who have ever lived.

Asymmetry of time between hearing and vision make it possible to do some tasks more easily with one sense than with another. Perception of sounds is much more closely tied to the time sequence in which they occur, while the time course of the perception of visual objects is often under the conscious control of the subject. Thus it is much easier to scan a book to obtain a particular fact than it is to scan a tape recording for a particular sound. On the other hand, hearing usually serves as a better warning of some change in environment, since a sudden sound always preempts attention, while the visual perception may be concentrated on some part of the visual field and ignore a change in some other part.

One of the prominent features of vision is color. Perceptually, one can inquire how similar or how different various colors appear. For example, experiments can be carried out by presenting pairs of colors to a subject and asking him to rate their difference on some sort of arbitrary scale, going from very similar to very different. R. N. Shepard [5] has developed a powerful technique for representing physical stimuli in a hypothetical perceptual space in which the distances between the stimuli in the space correspond to their similarity, objects which are close together being very similar and frequently confused, and objects which are far apart being very dissimilar and very infrequently confused. His technique has proved to be a most useful way of characterizing a wide range of stimuli from geometric designs to the quality of telephone circuits. Figure 11 shows the results of a similarity scaling applied to pure colors. The psychological space in which the colors exist could, of course, have various numbers of dimensions, three not necessarily being the maximum. It turns out that the colors can be well represented in

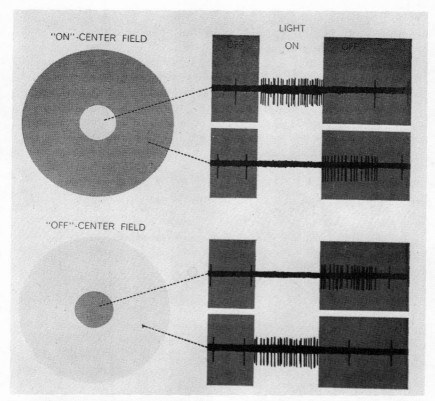

Fig. 12. Concentric fields characteristic of retinal cells. At top an oscilloscope recording shows strong firing by an "on-center" type of cell when a spot of light strikes the field center; if the spot hits an "off" area, the firing is suppressed until the light goes off. At bottom are responses of another cell of the "off-center" type. (From David H. Hubel, "The visual cortex of the brain," Scientific American, Vol. 209, No. 5, p. 57, 1963.)

a two-dimensional space as shown. The interesting thing about this representation is that it does not correspond to the frequency space which physically describes the colors. In the psychological space, the various hues form something approximating a circle, red being judged quite similar to violet, whereas in the physical space red and violet are at opposite ends of the spectrum, red being at the longest wavelength and violet the shortest.

Now let us turn to some physiological data which may give us some indication of how information is coded when it is transferred from the retina to the central brain. Most of the work described herein was done with cats by D. H. Hubel and T. N. Wiesel [6]. Various visual fields were applied to the eye of a cat, and the neural responses were measured by means of electrodes inserted at various parts of

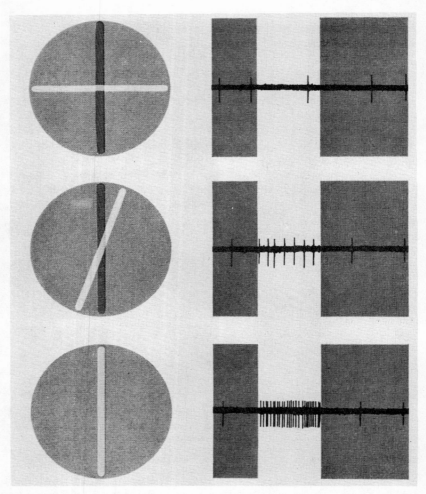

Fig. 13. Response of cortical cells sensitive to stimulus orientation. Horizontal slit (top) produces no response, slight tilt a weak response, vertical slit a vigorous response. (From David H. Hubel, "The visual cortex of the brain," Scientific American, Vol. 209, No. 5, p. 58, Nov. 1963.)

the brain.   Even if one places electrodes in the retina close to the rods and cones, very complex phenomena are observed. For example, as shown in Figure 12, there are both on-center fields and off-center fields.   In the "on" field, a spot of light at the center of the field produces pulses in the associated neuron, whereas in the off-center field, a spot of light inhibits pulses in the associated neuron and indeed pulses occur when the light is removed. But this is only the beginning of the complexity.   Surrounding the on-center field is an inhibitory area which when illuminated will inhibit the activity associated with the center of the field. Correspondingly, surrounding the off-center field is a facilitory area which will produce activity in the "off" neuron. The inhibitory effect is similar to that found by Frishkopf in the auditory neurons, and one can easily imagine that it is used to sharpen the contrast of lines in the field.

When one moves the electrode to the visual cortex, the observed phenomena become even more complicated. For example, cells can be found which respond only to a slit of light and, furthermore, only when the slit has a particular orientation. Thus, as shown in Figure 13, when the orientation of the visual field is vertical and the angle of illumination in the top section is horizontal very little activity occurs. As the orientation of the slit of light is rotated, the neural activity increases, and when it is coincident with the field orientation, a maximum activity is observed. By very painstaking work, Hubel and Wiesel were able to study, to an extent, the organization of the visual cortex with respect to the direction sensitivity

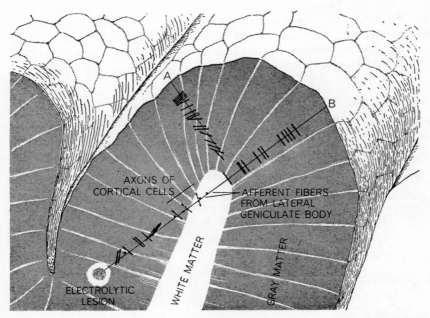

Fig. 14. Functional arrangement of cells in the visual cortex which resemble columns. Lines A and B show paths of two microelectrode penetrations; short intersecting lines show receptive-field orientations encountered. Cells in a single column had same orientation; change of orientation showed new column. (From David H. Hubel, "The visual cortex of the brain," Scientific American, Vol. 209, No. 5, p. 62, Nov. 1963.)

of the fields. This organization is indicated in Figure 14. The cortex is arranged in columns and the cells within a given column are all sensitive to the same orientation of the slit on the retina. Different columns are sensitive to different orientations.

I suspect that these few results, although they were produced by a great deal of excellent and painstaking work, represent only a bare beginning, and that we shall eventually find a whole Pandora's box of astounding mappings and functions between the visual image on the retina and the visual cortex. The complexity of the psychology of vision leaves no doubt that such must be the case, though there is much doubt that we human physiologists working in a very difficult area, where the best measuring tools are none too good, will be able to unravel this complexity. However, it is an interesting problem.

I would like to conclude with the comment that an adequate sensory input is absolutely vital to the well-being of any human. Certain unpleasant experiments in sensory deprivation have been performed in which the sensory inputs to a person are removed as completely as possible by bandaging him in a soft material, shutting out light and sound, and suspending him in a tank of neutral temperature water. These experiments have shown that he cannot and will not tolerate such a state for long. Vision and hearing are vital for our ability to communicate and seem intimately associated with our highest mental activities. We have only begun to understand the characteristics of these senses.

REFERENCES

1. Békésy, G. von., Experiments in Hearing. McGraw-Hill, New York, 1960.
2. Frishkopf, L. S., "Exitation and inhibition of primary auditory neurons in the little brown bat," J. Acoust. Soc. Am., Vol. 36, p. 1016 (A), 1964.
3. Hanson, R. L., "Clues to sound localization," J. Acoust. Soc. Am., Vol. 31, p. 1584 (A), 1959.
   Hanson, R. L., "Sound localization," J. Acoust. Soc. Am., Vol. 31, p. 830 (A), 1959.
4. Harris, Gerard G., "Binaural interactions of impulsive stimuli and pure tones," J. Acoust. Soc. Am., Vol. 32, pp. 685-692, 1960.
5. Shepard, R. N., "The analysis of proximities: Multidimensional scaling with an unknown distance function," Psychometrika, Vol. 27, Part I: pp. 125-140, Part II: pp.219-246, 1962.
6. Hubel, D. H., and Wiesel, T. N., J. Physiol., Vol. 160, p. 106, 1962.

The Confined Ambient—The Dirty Nest

Don D. Irish

Biochemical Research Laboratories
Dow Chemical Company
Midland, Michigan

In Central Michigan, there is a wild area we call the Dead Stream Swamp. On a warm, pleasant day in early summer, I sat in a bird blind watching an osprey feed its young. The adults cleaned every scrap of fish from the nest. The young ejected their fecal matter cleanly out of the nest and into the swamp. How cleverly nature had arranged to keep the nest clean. Musing on man and his environment, I contemplated the problem which would arise if the entire Dead Stream Swamp were paved with osprey nests, side by side and even vertically in a multistory structure. This in contrast to one osprey nest in thousands of acres of swamp.

The nest in which I was raised was a farmhouse in Maine. The old wood stove leaked at the joints; the acrid smell of wood smoke often filled the kitchen. We did not mind; the smoke meant warmth and food. The old farmhouse was full of cracks, and plenty of air came in and plenty of smoke went out. Most of it went up the chimney. Outside there was all God's outdoors into which the smoke could dissipate.

But times change. Houses are put side by side and stacked vertically. Wood is replaced by soft coal. In the 1930's, in one of our major cities on a day a native called "clear," I looked the sun straight in the eye; it was a dull, red, somewhat fuzzy spot. I am glad to say that on a recent visit the same city was much brighter and cleaner. Something can be done.

In my brother's diary of about 1905, at infrequent intervals was the announcement of a great event: "An automobile went by today." Yes, these early gas buggies smelled even worse than our cars today, but we did not see them often, and there was lots of space. Observing Chicago traffic, I decided times had changed.

As a young boy I used to love to swim in those clear, cold lakes in the state of Maine. Last summer I took a plunge into one of our more beautiful inland lakes in Michigan. As I came up to the surface, I felt a greasy film on my skin and looked around me to see an oil slick from the outboard motor that just passed. Times have changed.

Whose fault? Whose responsibility? Should the pot call the kettle black? Government? Did you ever drive by a burning city dump? Industry? The stack effluent is not always "pure steam." Other contributors? Our backyard incinerators, that smoke plume from the Diesel truck, the exhaust from many automobiles, the home furnace, many of the activities in our homes, at our places of work? Yes, everybody seems to be involved.

It is ironic, but it seems that about every time I attend a conference on air pollution, some individual carrying a big black cigar leans over to me, and as he puffs a choking blast of smoke in my face, he comments, "Can you imagine a man who would pollute the atmosphere that his neighbor must breathe?" Well, yes, I guess I can. Then he continues, "There should be a law against it." And I have to admit that there are times when I do feel that way. Of course, the law is supposed to apply to the *other* fellow.

I have included the outdoors in my illustration. However, please note the osprey cleaned its nest by ejection out of the nest. The wood smoke originated in the kitchen stove, the plume of "not-quite-so-pure steam" originated in the industrial operations. When we make a mess indoors, the easy solution is to throw it outside and let Professor Haagen-Smit take care of it. Yet the air which we use indoors comes from outdoors. On that farm in Maine, the vast open country could easily accept the little plume of wood smoke. We are now finding that the outdoors is not unlimited in its capacity to accept our enormously increased load. When I consider the distinction between the problem indoors and the problem outdoors, it reminds me of the Mobius strip, a toy of the mathematician. It is not only difficult to tell which side is the inside and which is the outside, but the outside becomes the inside and the inside becomes the outside.

Man's environment is the resource for all of his needs and the acceptor for all of his wastes. But this environment is neither an infinite resource nor does it have an infinite capacity as an acceptor. Our problem, then, is people and the activity of people in a not unlimited space. This problem increases as we have more and more people and more and more activity. What do we do about it? Just sit around and blow smoke in each other's face? Blame the other fellow and try to pass laws? Against the other fellow, of course.

Lest someone think that I am weeping for the "good old days," let me assure you that I don't want to go back to harnessing a horse. I'd rather step on the starter, particularly in near-zero weather. I don't want to go back to the old wood stove or lugging the wood it "eats." I like my house with automatic gas furnace and air conditioning. The harsher life did tend to thin out the weaklings and improve the race by this natural selection, but I would rather not be eliminated. I like the comforts of our modern age and I'll even take a little dirt with it if I have to, but I see no reason why we cannot do a cleaner job. While we are learning how to do a cleaner job, I am going to enjoy what we have.

New and useful materials needed in our modern society are being developed. We need to know how to handle them. They should be as appropriate for their designed use as we can make them at this time, and when knowledge allows we should make them even more appropriate. We need to be ready to meet the challenge of more people and more activity. It can be done.

Let us examine the logical steps to be used in attacking our problem. First, *to know*, as best we can, the potentiality for reaction between man and his environment. Second, *to communicate* this information as effectively as our present methods allow. Third, *to act* on this information, maintaining a healthful environment. It is sad to relate that sometimes we know but do not communicate, sometimes we act when we do not know, and sometimes we communicate what we think we know when it isn't so.

To know begins in the laboratory, extends to the work place, to medical research, and to the clinic. Let us start in one of our laboratories in Midland. While our goal is the maintenance of a healthful environment for man, our first approach must be animal investigations in the experimental laboratory. In our laboratory, we use many species of animals as this gives us a much better perspective of the possible modes of response. We use mice, rats, guinea pigs, rabbits, cats, dogs, men, and monkeys. When we need information on farm animals, plants, or insects, we turn to the agricultural people at our Bioproducts Research Center. When we are concerned with wildlife, we may study any of several of our native birds, animals, or fish. At times we may bring in almost any other type of creature that walks, creeps, crawls, flies, swims, or wiggles, depending upon the needs of our research. Several of these species we breed in our own laboratory, in order to assure ourselves of the consistent quality of our test specimen. Our strain of rats has been maintained in our laboratory for over 25 years, and while it may not be the most perfect experimental animal, we are certainly acquainted with most of its vagaries. Our 1964 crop of rats will number 40,000. Man we must take as we find him, not only with his individual genetic peculiarities, but also with his variable off-the-job habits and indulgences.

There are two facets to our investigations. First, to design exposures of animals which will be pertinent to the kind of contact that can occur in the manufacture, handling, and use of the materials under consideration, and secondly, the measure of response to a specific exposure.

Our first consideration is the acute handling problem, such as a spill, a break, or an accident. Does it injure on contact with the skin, when taken by mouth, when breathed as an air dispersion, or when splashed in the eyes?

As a material progresses in development, and its potential use becomes better defined, our toxicological investigations become comparably more sophisticated. The types of exposure to be expected in the projected use are more thoroughly investigated. If a material is to be used where skin contact is a major problem, appropriate skin studies are made. If a material may become part of the diet, it is put into the food of several large groups of animals, and they may be studied for a lifetime or several generations. If a material may be contacted as an air dispersion, known concentrations are maintained in a chamber, and several species of animals are exposed continuously or repeatedly for weeks or months, depending upon what is appropriate to the projected use of the material. This is most pertinent to our present consideration of air indoors—"the confined ambient."

Now that we have exposed animals, we must measure the interaction and the response to this exposure. The first interaction is at the molecular level. We study the absorption, distribution, metabolism, and the excretion of the substance or its metabolites. The second level of interaction is cellular. Cellular reactions culminate in a tissue response which is commonly designated as histopathology. Next is the level of organs or organ systems which is more commonly designated as pharmacology. Last and probably most significant of all, is the integration of these previous reactions into the whole animal response. There are many facets to the measurement of the whole animal response such as growth, physiological integrity or capacity, and behavior.

From this first step of "getting to know" in the laboratory, we derive an understanding of the quantitative and qualitative reaction of animals to the substance in question. Projection of this information to the probable reaction of man is our next responsibility. Such projections can be made with a rather high probability of practical usefulness. We should however never assume that we have the answer at this stage. From the information we have attained, we make a tentative estimate of the tolerable exposure.

Let us move to the Medical Research Laboratory. Here we may study ourselves in a very carefully-controlled concentration of known materials. Tests can be made of manual dexterity or of coordination and balance. Good biochemical measures of absorption and excretion in man can be established and compared to comparable measures in animals. Such measures are valuable tools in the clinic where men can be studied at a time when they are working in operations of manufacture, handling, and use. Our Medical Research Laboratory has made very effective use of some of the newer and highly-sensitive analytical tools to study expired air. Samples can be taken by blowing up a plastic bag. The air from the bag can then be analyzed by infra-red spectroscopy or by vapor-phase chomatography. They are able to pick up small amounts of material in expired air. By carefully-controlled experiments, they can find the relationship between time-concentration of exposure, and time-concentration in expired air. Knowing this relationship, when a man comes into the clinic, analysis of his expired air can give us some qualitative and quantitative clues to what his exposure may have been.

The next step in our effort "to know" is environmental research. As we have given a tentative judgement as to what is tolerable, we must be sure that the work circumstances, even in the early pilot-plant development of a new material, fall within these projected levels. The careful evaluation of the actual exposures, coordinated with good clinical investigations of the people involved, is another step in check-ing the validity of our projection from animal toxicological information.

Our environmental investigations are a means of determining a five-dimensional pattern: three dimensions of space, a fourth dimension of time, and a fifth dimension of man's movement within space—time. This is our best definition of the probability of exposure in a given set of circumstances. This concentration—space—time—man movement pattern is not a constant but a highly variable circumstance. In order to convert this into something usable, we often use multipoint analyzing, continuous-recording instruments. These are helpful in the daily control of the air in workplaces. For research purposes we may wish to integrate this data and it may be greater than can be readily handled by manual calculation. In some cases, for research purposes we have attached a digitizer to the continuous analyzing equipment, converting the data to punched tape which then can be mathematically integrated in our computer. We do not yet know how such data can be integrated to give us the most significant and useful information. We are recording analytical data, and relating it to the "state of health" of the people in the environment being measured. With luck we may be able to improve our understanding of the significance of these variable exposures.

To be of value, these measurements of exposure must be related to the "state of health." This raises the problem of defining the "state of health." However, the physicians in the clinic are developing newer tools which are making the assessment more definitive. Relating the "state of health" to exposure introduces other problems. If we are concerned with a work environment, we can make investigations of the work circumstance, but this represents less than a third of man's environmental exposure. What of his "off-the-job" habits and indulgences? In area outside industry, the facets of the environment are manifold, and their relation to health more tenuous. The same principles apply but the problems of application are more complex.

Now our second step is *to communicate* this knowledge and experience of the interactions of animals and man with the environment. This is easier said than done. What the other fellow thinks he hears is not always what we thought we said.

The first channel of communication is scientific publications which primarily informs one's scientific colleagues. Product brochures and labels are the most direct channels to the user, as they go with the product.

In the area of occupational health, the standards of environmental quality are a useful means of expressing and communicating the summation of experimental information and experience. These standards are given as threshold limits or acceptable concentrations. The Z37 Committee of the ASA is developing a number of innovations which I would like to briefly outline for you.

In the past, a standard was conceived as a single limit, threshold, or boundary between what is healthful and what is not healthful, but there is no such single boundary. What are the different acceptable concentrations most useful in environmental control? The following are at present under consideration by the Z37 Committee:

First, an "emergency exposure limit" for *a single event* which is limited to a level and duration which will allow rescue with no irreversible injury.

Second, an acceptable ceiling concentration for protection of health assuming eight-hour-per-day exposure.

Third, an acceptable eight-hour-per-day, time-weighted average within the limits of a specified ceiling and defined peaks beyond the ceiling.

Fourth, an acceptable maximum for peaks above the acceptable ceiling for continued exposure. There are many substances for which such excursions would be quite acceptable if properly defined.

Fifth, a minimum level for sensory detection qualified as to tolerance (sensory fatigue) and for warning value.

Finally, any other level or levels considered by experts to be pertinent to a particular substance under consideration.

The standard for each substance should contain sufficient information to clearly indicate the basis upon which each level is determined, the consequences of overexposure, and the relative adequacy of the information. Guesses by experienced people are useful until we have better information, but we should call them guesses.

I think it is obvious that we need to be constantly searching for a better understanding of man's response to his environment. Our past standards were very effective when competently used. We are proposing changes made possible by our newer knowledge of toxicology, yet we must be ready to change them again tomorrow, and again the day after.

We have now investigated the potential for interaction between man and his environment. We have attempted to exchange our information and experience so that what you have and what I have can be put together into a better overall understanding. From this we have attempted to express some reasonable and usable standards of quality. Now we can turn to our third step which is *to act:* to control the environment in our immediate operations and to develop more appropriate operations and products.

Within the workplace, we turn again to environmental research. We determine potential exposure in terms of space—time—man movement. We maintain medical surveillance, searching for biochemical and physiological signs of exposure. By operational control we can stay well within the environmental quality indicated by our standards. We also are on the watch for unforeseen problems and accidental aberrations.

As materials move beyond the work environment of manufacturing operations, control is less direct. Handling and transport are usually well-organized operations although communications are more fragmented.

As a material reaches the user in his home, our lines of control and communication are most tenuous. We must depend heavily on the label. We are tempted to add more information to make the label more complete. Although this may be good technically and legally, it may well defeat our main purpose, which is aiding the user. The longer and more complicated the label, the less probability it will be read or understood. It is too rarely read or understood in any case. We should be constantly on the alert for better means of communication with, and education of, the user. Our problem is not use but misuse; appropriate use comes with understanding.

So far we have considered our immediate and largely defensive action. This is critically necessary and desirable, but what are the possibilities of a more farsighted, a more creative approach? We must be constantly looking for more appropriate operations and more appropriate materials to fulfill the needs of our ever expanding society. To illustrate, let us take another look at the oil slick on my beautiful

Michigan lake.   One method of attack would be to pass a law to prohibit outboard motors. A law which must prohibit is an indication of failure. The constructive approach is to develop a water-dispersible, biodegradable lubricant.   No oil slick.   My friends can get to their fishing grounds and find more and cleaner fish, and I can enjoy a cleaner lake. This is constructive advance, in contrast to restrictive retreat.

Let us continue to view our problems from a similarly constructive viewpoint.   What a challenge to our scientists and inventors—a constructive approach to automobile exhaust, to the industrial stack, to my home chimney.   The answers are not all there simply waiting to be used.   We might develop atomic energy for all of our power, and the resultant electricity would heat our homes and run our cars. This will take time and a lot of hard work.

Did I hear a question?   Oh!   My friend with his cigar.   That is simple.   We can apply the usual industrial-hygiene practice, ventilate him "at the source," in other words, put a hood over his cigar and dump the effluent with all the rest.

We cleaned the nest, didn't we?   It's now in "the troubled outdoors," and, until he blows the smoke back into our nest, let's just turn it over to Professor Haagen-Smit.   Yes, I fear we too often tend to clean our nest and leave the effluent for the other fellow, then complain when it blows back into the nest.

Atmospheric Ecology II:
The Troubled Outdoors

Arie J. Haagen-Smit

Department of Biochemistry
California Institute of Technology
Pasadena, California

The discoverer of the essential fraction of air, a young minister of the Dissenting Church in England, Joseph Priestley, wrote as follows:

"My reader will not wonder that after having ascertained the superior goodness of dephlogisticated air by mice living in it I should have the curiosity to taste it myself. I have gratified that curiosity by breathing it. The feeling of it to my lungs was not sensibly different from that of common air but I fancied that my breast felt peculiarly light and easy for some time afterwards. Who can tell but that in time this pure air may become a fashionable article in luxury. Hitherto only two mice and myself have had the privilege of breathing it."

Priestley did foresee its use in hospitals in cases of shortness of breath when he writes:

"It may be peculiarly salutary to the lungs in certain morbid cases when the common air would not be sufficient to carry off the *phlogistic putrid effluvium* fast enough."

He may well have felt its need when breathing the polluted air in his hometown of Birmingham. This industrial center matched London with its black fogs, which Chateaubriand upon his arrival in England described as follows:

"Soon I saw before me the black skullcap which covers the city of London. Plunging into the gulf of black mist as if into the mouths of Tartarus I crossed the town."

This was the year 1822. Under this black skullcap happened the mass poisoning of 1952 when several thousand excess deaths were reported in a four-day lasting black fog. During much of this time visibility seldom exceeded 100 yards, and in some parts of the city it was considerably less and even zero has been reported. Consulting the newspapers of those days, December 6 to 9, is somewhat disappointing. There are no four-inch-lettered headlines, but mention is made of the complete dislocation of transport by three days of fog, and one has to turn many pages of *The Times* to collect items which have some bearing on this much-quoted disaster:

Excerpts from *The Times*

Monday, December 8, 1962		Tuesday, December 9, 1952
TRANSPORT DISLOCATED BY THREE DAYS OF FOG		MORE DEATHS AT CATTLE SHOW
No Improvement Expected Today	Unbroken Frost	London Queue of 3,000
Visibility "Nil"	Sunny Periods on Coast	Many Road Crashes
London Buses Stopped	Street Attacks in Fog	Roundabout Closed
Cars Abandoned	Housebreakers Active	Aircraft Diverted
Thick Weather Watch	Post Office Robbery	Sunshine on Coast
Road Accidents	Many Sports Events Postponed	
River Traffic Halted	Antwerp Paralyzed	
Show Cattle Affected		

One would expect that after many of these smoke attacks, the Londoner might lose his patience and insist on corrective action. For such a move, the help of the press is essential, but they apparently did not see anything unusual in the black fog and instead we find a touching editorial in *The Times* on December 9, 1952, on the virtues of the fogs from the ever-present defender of the *status quo*. From his pen came these reassuring words:

"Fog in the Fields: .... The truth that fogs have taken to rubbing into us of late years is that they are not the parasites of coal fires and other dirt creating human agencies they were once accused of being. ... The fogs are ancient Britons. They met the boat when the ancestors of Boadicea landed. ... Robbed of the full glory of brown and yellow they can still mix on their old world palettes a striking study in grey. ... The country side dissolves under the spell into a parody of fairyland. Cattle turn into ghostly dragons, breathing out for those who came near them a stream that is more eerie than fire. ... Noting all in the interval of stumbling to the byres the farmer, joining his wife who has been feeling her way from the fowl run, hears news of the great city with a comradely smile. The fog has not forgotten to pay him a visit."

One feels warmed by such statements and I would forgive the

author if it were not for the fact that when the score became known there were 4000 deaths, most of the fatalities occurring among the population from age 45 upward. The time lag between the arrival of the smog and the beginning of increased mortality was less than one day. During the period of several days when the overall level of pollution was on the increase, the number of deaths rose corresponding-ly, and when the pollution decreased, the death rate declined also.

It is quite clear from reading the newspaper reports, that there was little realization of imminent danger during the pollution period. The mortality remained elevated for several weeks after the weather had improved and the extent of the disaster became known a consider-able time after the four-day period of black fog. A careful survey of past records has shown that several similar episodes have occurred in the past in England. In our country, we can point to Donora near Pittsburgh, when 43% of the population was made ill during the 1948 pollution period, and when 20 deaths occurred which could be directly attributed to the severe pollution of the air. Of special interest is the fact that a follow-up study of a previous episode revealed that those affected but recovered in 1948, did have 10 years later a less favorable mortality and morbidity rate experience than those who were not affected during the pollution period.

Other warnings have been given to us that the air is not unlimited, and that it is possible to pollute it so much that widespread irritating and even toxic effects begin to show up. We can add to the list of pollution incidents by pointing to the large-scale respiratory difficulties in New Orleans and Yokohama, and the atmospheric poisoning of the population of Poza-Rica in Mexico. Of special interest are also the studies by Dr. Greenburg and his colleagues, which showed that New York City had a history of pronounced higher death rates during severe air pollution periods.

Typical for all these cases is that the analytical findings on the concentration of the pollutants are usually well below accepted health standards, and this should be a warning that community health stand-ards cannot easily be too stringent. However tragic these episodes are, they are soon forgotten in competition with 40,000 deaths per year in automobile accidents, and the many thousands of violent deaths from other causes. It was especially the plight of Los Angeles which directed world-wide attention to the need for the preservation of the quality of our air. There is an important difference between the earlier air pollution disasters and Los Angeles smog. Los Angeles smog is not easily forgotten because it occurs about 200 days of the year, and affects in some way or other about 80% of the people. This means that smog is with us most of the time, and it is not the other fellow that suffers. Nearly everybody is aware of some of the symptoms.   Someone may not be sensitive to the eye irritants, but he may not like the chemical odors of smog, or he may object to the strong haze which usually appears when smog is at its worst.

The farmer is hurt when his crop has become unsalable because of leaf damage from smog. Rubber goods are more severely attacked by smog air than by normal air, and this is true for many other materials. No one really can make an estimate on this type of damage, but it is said to run into billions of dollars for the United States alone.

It is quite normal to see concentrations of carbon monoxide rise to 10 and 15 parts per million and in the heavily traveled streets concentrations of 50 and more are registered. Ozone concentrations often reach levels which are of definite concern to the medical profession. When the smog haze comes in with its typical odor, one can see the damage appear on the leaves of sensitive plants. Observing this damage is one of the most convincing demonstrations of the toxic effects on living tissues. After a few days the affected parts die and a damage pattern is formed typical for the action of the pollutant. Observation of this damage pattern in different plants has become a convenient and inexpensive indicator of smog, which has been used extensively by plant physiologists in establishing that smog is not limited to Los Angeles, but is widespread both in the United States as well as in other parts of the world.

Photochemical air pollution injury to vegetation now occurs in 25 countries of California distributed in the south coastal basin of Los Angeles and San Diego, the San Francisco Bay Area, and the San Joaquin and lower Sacramento portions of the great Central Valley. Oxidant damage to plants has been reported from a number of European and South American cities as well as in the United States at Baltimore, Maryland, Philadelphia, Pennsylvania, and New York City. Typical oxidant plant damage now is reported in the western states of Washington, Utah, and Colorado, as well as in northwest coastal Baja California, Mexico. Similar damage has been observed in the midwestern states of Missouri and Illinois. Although oxidant injury occurs in many eastern states, ozone damage seems to be predominant in Connecticut, New York, New Jersey, Pennsylvania, Delaware, Maryland, District of Columbia, and North Carolina, and heavy damage to the tobacco leaves has been reported.

Apparently something has happened to our air. It is not the life-giving oxygen which Priestley mentioned which has changed. Even though a tremendous amount of oxygen is consumed in fuel burning, the available oxygen concentration would at worst decrease by only 0.01%. The carbon dioxide formed in these combustion processes does measurably increase the concentration in the air from its 350 parts per million with about 0.07 parts per million per year. This is not an insignificant increase over a long period of time, and speculations have been made about its effect on the heat balance of the earth. However, this increase, in our lifetime and a few generations to come, is not contributing to any observable ill effect, since we tolerate relatively large fluctuations in the carbon dioxide content of the air.

The reason for the observed deterioration of the air is the addition of trace substances. Some of these substances measure not more than a few tenths of a part per million. Orchid flowers are severely damaged by concentrations of 5 parts per billion of the automobile exhaust component, ethylene. Fluoride damage to the leaves of gladioli is noticeable at even greater dilution in the order of a tenth of a part per billion.

The sources of these contaminants are generally well known and originate from incomplete combustion of various fuels, from various industrial processes, and from miscellaneous operations such as trash and damp burning. To answer the question, why do we pollute the air in this irresponsible manner, we can go back to another scientist of a few centuries ago. Robert Boyle, well known for his formulation of the gas laws, said: "The generality of men are so accustomed to judge of things by their senses, because the air is invisible, they ascribe but little to it and think it but one remove from nothing."

To most people, air and space are about synonomous; both seem to be available in unlimited amounts to please mankind. Our generation has discovered that unfortunately nothing on earth is unlimited, and we have begun to worry about the limitation of our soil, our water, and, lately, our air. Some of us feel that our generation has a responsibility to future ones in using these resources wisely. It is clear now that we cannot any longer use the air indiscriminately as a dumping ground of all our by-products, especially when the natural ventilation is inadequate. This lack of ventilation is a prime factor in the formation of the Los Angeles smog. Mountain ranges prevent unobstructed horizontal flow of the air, while frequent temperature-inversion conditions limit the vertical movement.

Pollutants trapped under the inversion layer move back and forth under alternating sea and nightly land winds. Tetroons, little weather ballons used by the U. S. Meteorological Service, show clearly the complicated path of the air packages under these conditions. Trajectories determined by radar response from the balloons show that in one day the air over Long Beach may reach halfway to Catalina Island, to visit upon its return Santa Monica, Los Angeles, and Pasadena. The conclusion we can draw from such studies is that pollutants released in one area of the basin are likely to blanket the whole area.

Simple arithmetic shows that under inversion conditions of a thousand feet, only a few hundred tons of chemicals suffice to establish concentrations of several tenths of a part per million. The main source of these irritating substances is a photochemical oxidation of gasoline components in the presence of oxides of nitrogen. This oxidation gives rise to high concentrations of ozone, and some of the intermediate reaction products have been shown to produce the irritating properties usually associated with smog.

Although the symptoms of photochemical or Los Angeles-type smog, which is mainly caused by automobile exhaust, are appearing in other cities, many of the industrial areas are still plagued by the classical forms of air pollution — smoke and fumes accompanied by gases such as sulfur dioxide.

It is a fallacy to think that conditions favorable to high concentrations of air pollutants are limited to the West; in fact, this particular meteorological condition is quite frequent over the whole country. Everyone is familiar with the morning inversions, when a cold layer of misty air remains close to the ground. An airplane trip across the country shows quite clearly the horizontal expansion of the smoke plumes covering hundreds of square miles.

A spectacular picture of the progression of a pollution cloud was observed by Dr. Davis, who photographed the progression of the smoke from the big Hancock fire of 1958 in the Los Angeles basin. The hot gases rose to 4000—5000 ft, after which they spread abruptly in a horizontal direction toward the northwest. Even though ground wind direction would not have predicted that the smoke would reach Pasadena 25 miles to the north, in a very short while soot started to fall in that area, and millimeter-large pieces settled down on the 100 in. telescope mirror of Mount Wilson.

For this reason one should think of airsheds as natural units of air pollution, just as in the past we have been discussing water pollution problems on the basis of watersheds which are determined by natural boundaries rather than political ones. Such coherent areas of air pollution are easily identifiable. For example, the Bay Area, comprising San Francisco, Oakland, San Jose, and other sizable cities, forms one unit as far as pollution is concerned. On the Eastern seaboard, the area from Boston to New York to Richmond forms one unit, and in the Middle West the airshed stretches from Michigan across the border into Canada.

The increase in population in these airsheds makes it imperative that legal structures be developed which exceed the boundaries of the local agency. In the Southern California airshed, the controlling agencies are in the hands of the individual counties, which have a voluntary nonofficial working agreement toward clearing the air.

Such agreements will be honored as long as the economic pressures are not too strong; eventually, however, they have to be replaced by more formal legally binding compacts.

The control districts have formulated rules limiting emissions, such as weight of particulate matter, or of sulfur dioxide emission. In a few cases the use of potential polluting materials have been prohibited unless they satisfied standards set by the control agency. Examples of this are found in the regulations on the burning of soft coal in the Midwest and the burning of residual oil in power plants. An important function of the control agency is to see to it that building of new industrial facilities incorporates all the necessary air-pollution equipment.

It is also a task of the agency to monitor the area for contaminants so that during a dangerous increase of pollutants sufficient warning can be given and preventive measures taken. The monitoring system has played an important role in the setting of standards for reasonably acceptable air. In the course of many years of measurements of the oxidizing property of Los Angeles air, it was shown that at about 0.15 part per million of oxidant, as measured by the liberation of iodine from potassium iodide, eye irritation begins to be noticed by a substantial part of the population. Quite often this oxidant value is materially exceeded, and it was decided that at least an 80% reduction of the incompletely burned automobile exhaust gases would be necessary to reduce the irritation to a few days per year. On the basis of this judgment, certain criteria were adopted for corrective devices on automobiles. To implement these standards, a State Motor Vehicle Board was created which was assigned the task to certify control devices developed by the industry.

Considerable progress has already been made. One point of emission, that of the crankcase, has been corrected in all new cars by a system whereby the gases from the crankcase are returned to the cylinders where they are burned.

For the larger contribution of partially-burned fuel, important progress can also be recorded. A system developed by Chrysler finds a solution in burning a leaner air—fuel mixture combined with certain changes in carburetor and spark timing. Other companies such as Ford and General Motors have concentrated on burning the excess fuel in the exhaust manifold by introducing additional air. Such systems will be available on the cars sold in Los Angeles in September, 1965. It is most likely that within a few years all new cars in this country will be equipped with these or similar devices.

The sad experience of Los Angeles promises now to be of great help to other urban areas. There is, however, no reason to relax. The special meteorological conditions described earlier for Los

Angeles require that we continue attempts to clear the air of any objectionable and foreign materials. This is necessary because our cities are growing at a rapid rate. Los Angeles increases every year by about 200,000 people. They bring in 100,000 cars and their industry, and any control measure has to cope with this increase. It has been a wise decision of the U. S. Department of Health, Education, and Welfare to invest considerable funds in establishing professorships at various universities to educate men who can cope with the infinite number or problems of modern air pollution.

From these centers of study, we expect to come refinement in engineering methods, which can offset the increase of pollution due to the expansion of our cities. I also expect close liaison of these front-line men with those in other fields of conservation. Keeping the air fit to breathe is only one aspect of wise management of our natural resources. Everyone studying these problems comes to the same conclusion — that fixing up cars and industries with control equipment is necessary but that the relief is temporary. A more permanent solution can be promised when it is more generally realized that air-pollution problems are intimately connected with the development of our community, and the application of our skills in engineering and science are related to problems of community living.

At present it is neither our lack of engineering know-how which is holding up the control of air pollution in this country, nor is it that we are too poor to afford the money which is needed for these controls. The difficulties are inherent in government of people; we have to convey to a sufficiently large number of citizens that air pollution is bad from the point of view of health, and costly from a purely economic point of view, and that it does interfere with the proper enjoyment of life in general. The hundreds of autonomous, largely self-serving structures which make up our large urban developments need to work together wholeheartedly among themselves and with state and federal authorities to master the insidious contamination of the air.

When an enemy tries to overrun this country, we seem to be able to get together and set narrow privileges aside for a little while until victory has been won. Why not do the same in fighting the enemy within our midst? And don't forget that the forces which tend to destroy our natural resources are many, and with the rapidly growing population they will be even stronger in the future. It is up to those who recognize the value of these resources to band together in programs which will spread the word — that conservation of all our resources is essential to preserve our way of living.

Atmospheric Ecology III:
Remedial Atmospheres

Capt. Albert R. Behnke, M.D.

United States Navy (Retired)
San Francisco, California

INTRODUCTION

Global fall-out, air pollution, and the prevalence of respiratory diseases are our bequest to the next generation. Persons thousands of miles from the source of detonation of a nuclear weapon have absorbed the short-lived iodine-131 and subsequently long-term, slow-decay fall-out products, such as strontium-90 and cesium-137. Problems of air pollution arise from heavy industrialization and the flagrant combustion of petroleum products. With respect to communicable disease, it has been shown in tests involving purposeful dispersion of noninfectious saprophytes, that thousands of miles of coastal area could be contaminated if the biologic agents were infectious.

Chronic obstructive pulmonary disease in the United States, less publicized than cardiovascular ailments, poses a formidable problem [1]. During the past five years death rates for diseases of the respiratory system have risen abruptly, and as in Britain, middle-aged men are the prime targets. It has been established in England that sickness and death from the complications of chronic bronchitis are highest in congested cities where air pollution is at its worst.

At the present time there is a renewed emphasis on the effects of meteorologic conditions on diseases and their therapy [2, 3]. There also has been forced upon us the need to define specifically the habitable submarine atmosphere and the optimal breathing medium in the space capsule. In response to the challenge of these problems, and as a result of the sophisticated biochemical and biophysical characterization of the atmosphere, climatology is undergoing a metamorphosis. There is an expanded knowledge of specific physico-chemical mechanisms underlying biologic response which enables the engineer to provide specified conditions of temperature, moisture content, and air movement.

The benefits obtained from remedial atmospheres may be grouped into three categories.   In the first, as exemplified by submarine operations, are striking and unequivocal benefits derived from control of temperature, humidity, and removal of pollution, chiefly chemical. Thus it has been possible to convert a previously noxious submarine atmosphere to acceptable metropolitan standards. Despite tremendous heat loads imposed by nuclear power, it has been possible to maintain a temperature of 70°F during a cruise under the Polar icecap and in tropical waters as well.   In our southern states the striking benefits obtained from air cooling of industrial plants, office buildings, homes, and even automobiles have erased the lethargy attributed to a debilitating climate.   In the second category are therapeutic benefits, not easily defined or quantified and not dependent upon removal of noxious agents from air, or control of temperature and humidity, which result from the addition of agents such as medicated aerosols in inhalation therapy, as well as air ions and ultraviolet light.   A third category centers in the renascence of hyperbaric therapy to provide, chiefly at present, oxygen at high pressures in the effort to relieve the acute asphyxia of cardiovascular emergencies, to treat specifically fulminating gas gangrene infection, and to provide increased tissue oxygenation in cardiac surgery and X-ray therapy.

WEATHER AND CLIMATE IN RELATION TO DISEASE

Seasonal Phenomena

Although countless persons afflicted with diseases, chiefly pulmonary tuberculosis, have sought out areas such as New Mexico and Arizona, a textbook of medicine today may not contain a single statement in regard to the beneficial effect of climate or to the historical relationship of weather and climatic conditions to respiratory ailments. Respiratory disease may show a fourfold increase in the winter months of a temperate climate in both the northern and southern hemispheres, whereas gastrointestinal ailments have a strikingly high incidence in the summer months. This seasonal phenomenon, however, cannot be attributed to any specific meteorologic parameter any more than one can state the reason why certain types of joint pain are predictive of rain.   It appears that weather and climate are intertwined with other factors to create an adverse environment. Thus, in winter months, atmospheric inclemency may lead to crowding of persons indoors without adequate ventilation.   There is generally overheating and the resultant low humidity dries protective secretions on congested mucous membranes.   Associated with weather conditions are polluted urban atmospheres, heavy in chemical and organic aerosols, which together with indigenous cigarette smoke irritate the respiratory tract.   In patients embarrassed by cardiovascular and respiratory ailments, the induced bronchospasm may precipitate respiratory embarrassment and lead to fatal sequelae. The seasonal influence is also apparent during the growth period which in such climates is more rapid in summer than in the winter.

Remedial Atmosphere Defined as an Atmosphere Free from Pollution

It is not in mortality or even morbidity but in the gray zone between fitness and illness that the effects of a smog atmosphere are striking. Allergic persons who descend in airplanes or in automobiles from the clean atmosphere of the summit of the Ridge Route into the smoke-filled basin of the Los Angeles area cease to enjoy a symptom-free condition, and within a period of several hours develop eye irritation, nasal congestion, sore throat, wheezing, cough, and subsequently disturbed sleep.   Apart from the undisputed role of pollutants and weather conditions in precipitating the untoward symptoms, and the fact that relief is obtained by return to a relatively clean atmosphere, it is not possible for the bioclimatologist to define positive physical characteristics and qualities of a remedial atmosphere.   Thus, the conclusions drawn from a seven-year study of possible relationships between climate, weather, and bronchial asthma not influenced by pollen count, are expressed only in general terms, namely, that the frequency of asthma was found to increase rapidly after a sudden increase in general turbulence of the atmosphere if combined with the influx of cold air masses.

The Optimal Climate

The physiologist has described the effects of extremes of heat, cold, and high altitude in quantitative terms.   The studies of Dr. Clarence Mills [4] have pointed out the debilitating effect on military men, for example, in abrupt transfer from a temperate to a tropical climate, and their lowered resistance to respiratory infection and sensitivity to cold on transfer from a tropical to a temperate or a subarctic climate. However, we do not find a description of the optimal climate, that is, the climate associated with the highest degree of fitness, productivity, and the lowest morbidity and mortality. Although it is not difficult to give a description of a deleterious atmosphere in biochemical and physical terms, a definition of optimal climate is complicated by the age and sex of the individual, and the type and degree of pathology affecting him.   The premature infant requires relatively high air temperatures and humidity, the healthy adolescent thrives on the coast of Maine, and the aged look to the sunshine and warmth of Florida. If one selects as favorable the climate of Bermuda, then unfavorable comparisons can be made in terms of constitutional vigor between the same descendants of Scottish-English stock who emigrated to Canada and those who emigrated to Bermuda. The disparity between the two groups is compounded in successive generations. Likewise in Hawaii, periodic leaves of absence appear to be obligatory for well-being.

Behavior and Weather

A considerable volume of descriptive literature characterizes the abnormal, unpredictable, and often bizarre behavior patterns of individuals and populations subjected to episodes of hot, oppressively humid weather. The hot winds blowing from the Libyan desert become moisture-laden over the Mediterranean and give rise to sirocco weather, which profoundly affects mood and behavior in the Southern European countries affected. This type of weather impairs heat loss from the body, interferes with rest and sleep, and is conducive to irritability and chronic fatigue. Striking is the irrational and at times criminal behavior manifest in individual and mob violence. Racial outbreaks in the United States are prone to occur on slight provocation during periods of hot, humid weather. Aboard ship it was found that cool living quarters conducive to rest and sleep without sweating prevented irritability, heat rash, and chronic fatigue which otherwise supervened. One may well reflect on the fact that the destructive violence of a single outbreak may exceed the cost of providing cool living quarters for a lifetime.

REDEMIAL ATMOSPHERES IN CLIMATIC CHAMBERS, SHIPBOARD SPACES, AND SUBMARINES

Climatic Chambers

Climatic chambers can be designed with the capability to reproduce or control any given climate. The factors involved in this control are air temperature, humidity, wind velocity, temperature of nearby surroundings, solar radiation, ionization, precipitation, barometric pressure, and provision for varying all of these in accordance with diurnal and seasonal fluctuations [5].

Of the various climatic and meteorologic factors, only a restricted number have been applied in climatotherapy. Investigation connected with the use of climatic chambers are arduous, require extensive supervision, and the results generally, outside of the field of temperature investigation, have not been commensurate with the scientific effort and expensive equipment required.

Low-Pressure Chamber

The physiologic effects of simulated high-altitude treatment (2000-3000 m, 6500-9850 ft) administered to asthmatic and bronchitic patients refractory to other forms of treatment have been reported

by Tromp [6].   Some 30 patients were given 11 to 119 treatments
(one-hour exposure, three to five times weekly) over a period of
two years.   Eight patients showed marked improvement, 10 moderate
improvement, and 12 slight improvement.   In addition to amelioration
of symptoms, it was possible to demonstrate improved pulmonary
function (measurement of forced inspiratory and expiratory volumes)
and improved thermoregulation following immersion of the hand in
cold water.   Yanda and Herschensohn [7] reported clinical benefit
to four emphysema patients from eight exposures each to an altitude
of 18,000 ft (oxygen).   The beneficial effects persisted for several
weeks.   The improvement brought about in the patients illustrates
the benefits possible in a remedial atmosphere which involves an
appreciable change in pressure and the possibility of thereby read-
justing deranged intrapulmonary pressure relationships. Noteworthy is
the persistence of improvement in chronic pulmonary disease, and
this feature will be pointed out later in dealing with hyperbaric therapy.
It is not possible, however, to summarize concisely the beneficial
effects obtained from the multifarious treatments accorded thousands
of patients.   Sufficient data from scientific studies are not available.

Submarine and Shipboard Compartments

It is difficult today to comprehend the impediments and opposition
to the air conditioning of ships in a bygone era. The philosophy
that military effectiveness would be reduced by displacement of an
extra gun with air-conditioning equipment has had serious consequences.   Military attitude toward conditioning of air in submarines
is succinctly stated as follows, "It has long been realized that
people perform far more efficiently under comfortable working conditions.   Men wilting from the heat and humidity inside the submarine
could not be expected to do as good a job as those same men envigorated
by cool, comfortable air.   This is a reason for air conditioning in the
submarine.   But the primary reason for air conditioning the submarine
is to reduce electrical grounds, specifically grounds on the main
storage batteries.   By air conditioning, moisture is removed from the
air, the air is drier, and moisture grounds on the main storage
battery are less likely to occur" [8].

It was possible years ago to obtain reliable data on watchstanders
during work.   The simple techniques employed can be applied to
assess atmospheric conditions on large numbers of persons.   Thus,
a shielded bulb thermometer placed over the chest between skin
and undergarment registered "envelope" temperatures which were
equivalent to the average skin temperature of the trunk of the body.
A special thermometer which could be worn continuously was placed
with its bulb inside the shoe adjacent to the instep of the foot.   The

temperature registered was close to foot temperature. In taking oral temperature, the bulb of the thermometer was placed under the tongue for five minutes. The temperature registered proved to be a satisfactory index of deep body temperature. In Table I pulse rate and temperature data are recorded on a watchstander standing duty in a "hot" space. He had periodic access to "spot" cooling provided by a blower in a duct that drew in outside air. It is apparent from the decrease in pulse rate that heart action improved during successive hours of the watch.

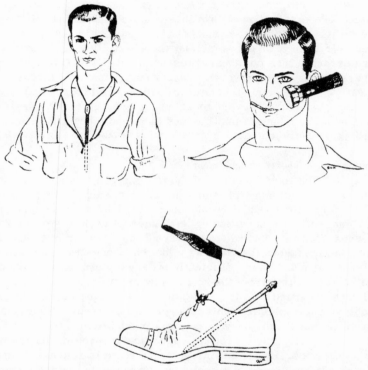

Fig. 1. Technique for recording observations on active persons in the effort to assess atmospheric conditions. Envelope temperature is measured by a thermometer worn continually, the bulb of which is shielded between shirt and undershirt and underlying skin over the chest. Oral temperature is observed by placing the thermometer under the tongue for a period of five minutes and read in place as shown in the illustration. Smoking, fluids, and food are prohibited for a period of one hour prior to reading. Foot-shoe temperature is obtained by placing a specially-constructed thermometer, kept in position continuously, so that the bulb is adjacent to the instep. This temperature is responsive to the temperature of the foot and of the shoe itself, which in turn is affected by the ambient air and the deck on which the men work. Condition of the skin of the forehead is recorded as either dry, clammy, wet, or dripping. Mental state is described as asleep, drowsy, awake, or alert. Subjective response to air conditions (after Yaglou) is recorded on a fixed sensation scale using only the following terms: cold, comfortably cool, comfortable, comfortably warm, and hot.

Table I. Temperature and Pulse-Rate Data (°F) Recorded on a Watchstander in the Engine Room of a Navy Cruiser in 1936 (Unpublished Data, A. R. Behnke)

Hour of watch	Oral temperature	Envelope temperature*	Foot−shoe temperature†	Pulse rate	Effective temperature‡
1	99.3	98.5	101.0	144	95.5
2	99.2	96.8	101.0	140	95.5
3	98.6	95.5	101.2	108	95.5
4	98.6	96.0	101.2	116	95.5

*Equivalent to skin temperature over the chest area. The bulb of the thermometer was adjacent to, but protected from, the skin by a perforated plastic shield.

†Temperature of the air in the shoe measured with a special thermometer inserted at the instep of the sole.

‡The temperature of the watchstander's environment, but out of range of "spot" cooling. The effective temperature equivalent to a wet-bulb temperature of 95.5°F did not incorporate a correction for radiant heat.

Under milder conditions, the same kind of data (Table II) are recorded as mean values on 10 men observed for three hours daily in a compartment of a warship en route from Lima, Peru, to Long Beach, California. The favorable response of blood pressure and pulse rate (taken in reclining and standing position to provide, in part, the Schneider index) is much higher (a favorable response) under cool than under warm conditions. In the tropical area, the effective temperature of 85°F was not conducive to sleep without sweating. When the compartment was air conditioned, the body and environmental temperatures were nearly identical with temperatures recorded in Long Beach under conditions of natural ventilation. Small environmental changes separate a condition of well-being from one of fatigue.

In Table III, pulse rate and temperature data are recorded on an entire crew which stood the same watches during surface cruising off New England and subsequently in the Caribbean area. When outside air was available for ventilation during surface cruising, there was no problem with respect to temperature and humidity.

From the physiologist's point of view the submarine is a metabolic chamber. Under no other circumstances is it possible to observe some 70 to 100 men under rigidly controlled conditions for many weeks and even months. Compartment temperatures in the older submarines were about 10°F higher than sea water temperature. In ships submerged in semitropical waters (sea temperature, 78 to 84°F), men sweat continuously. In a typical three-day cruise in tropical waters, the average weight (fluid) loss amounted to about five pounds per man. In the early thirties a submarine officer demonstrated that by diverting

Table II. Physiologic Observations of Navy Men in a Warship Compartment During a Cruise from Lima, Peru, to Long Beach, California, in 1936 (Data from A. R. Behnke) Mean Values for 10 Men

Hour of test	Schneider index*	Pulse rate (sitting)	Body temperature, °F		Air temperature, °F			Remarks
			Oral	Envelope†	dry-bulb	wet-bulb	Effective	
Day 1								
1	10.7	70	98.9	88.6	78.5	69	74.5	Lima, Peru
3	11.4	67	98.8	89.7				Men comfortably cool
Day 2								
1	8.8	75	99.1	91.9	83	73	78	Men comfortable or comfortably warm
4	9.9	68	98.9	92.1				Sea temperature 68°F
Day 3								
1	7.6	77	99.4	91.7	83	73	78	5 men comfortable and awake, 3 men comfortable and alert, 2 men comfortably warm
4	9.9	66	98.8	91.3				Sea temperature 70°F
Day 4								
1	8.1	74	99.2	–	87	80	83	Out of Humboldt Current.
4	9.5	68	98.8	92.6				Sea temperature 83°F
								6 men comfortably warm, foreheads damp, 4 men too warm, sweating

Day								
Day 5								
1	5.7	76	99.1	—	91	81	85	3 men hot, sweating, 4 men too warm, 2 men comfortably warm
4	8.1	68	99.1	93.6				
Day 6								
1	7.2	76	99.2	—	89	80.5	84	8 men comfortably warm, 2 men hot
4	8.5	69	99.0	93.6				
Day 7								
1	5.7	77	99.3	93.7	91	81	85	4 men hot, sweating, 3 men too warm, 2 men comfortably warm. Sea temperature 83°F
4	7.2	70	99.0	—				
Day 8								
1	8.7	73	98.9	93.1	87	76	81	Air conditioned Men comfortable or comfortably warm
4	11.0	65	98.8	91.6	83	65.5	75	
Day 9								
1	10.4	69	98.9	90.2	80	68	74.5	Long Beach, California Sea temperature 62°F Men comfortably cool
4	11.9	60	98.4	89.1				

*Based on pulse rate and blood pressure; the higher the score, the better the physical condition.
†Temperature of air layer between undershirt and chest.

Table III. Average Values of Temperatures and Pulse Rate Recorded on Watchstanders in a Submarine Operating off the Coast of Maine and Subsequently in the Carribean Sea, August and September 1940

Date and hour	Body temperature, °F			Pulse rate	Sea water temperature, °F	Compartment temperature, °F, and relative humidity		Remarks
	Oral	Envelope*	Foot-shoe†			Control room	"Hot" spaces‡	
						T – RH	T – RH	
Maine area								
8/22								
1000	98.3	90.8	87.7	78	60	82 – 58	–	Submerged; recirculation of air without cooling
1300	98.4	89.9	90.0	84	60	81 – 63	–	At 1700, recirculation and cooling
1700	97.8	87.4	87.2	77	60	80 – 59	–	
						Control room, and living spaces	Hot spaces	
Caribbean Sea								
9/17								
1000	98.5	93.3	96.0	82	84	90 – 67	109 – 47	Sumarine submerged; air recirculated Most men sweat during sleep; 4 developed headache; minimal air cooling
1300	98.7	92.5	96.1	83	84	90 – 67	105 – 45	
1700	98.6	93.5	96.0	81	83	90 – 64	103 – 50	
9/18								
1000	98.5	92.6	95.8	79	84	90 – 68	107 – 46	Over half of crew sweat during sleep; 4 men had headaches; minimal air cooling
1300	98.5	92.6	96.1	83	83	90 – 66	105 – 47	
1700	98.9	92.8	95.7	83	84	89 – 67	103 – 50	

						Control room	Maneuvering space	
9/19								
1000	98.5	93.0	96.1	82	84	90 − 69	105 − 48	Half of crew sweat during sleep; 8
1300	98.7	92.4	96.7	85	84	90 − 70	104 − 51	men developed headaches as the
1700	98.6	92.8	95.8	81	84	90 − 71	100 − 55	CO$_2$ rose above 2% minimal cooling of recirculated air
9/20								
1000	98.5	93.1	96.8	80	85	91 − 77	103 − 53	Reduction of cooling of recirculated air to one compressor unit
1300	98.8	93.6	97.1	90	85	92 − 78	103 − 59	Surfaced at 1440
0845	98.4	92.6	96.0	91	−	93 − 63	−	Submerged; compressors secured; recirculation of air − no cooling
0900	−	−	−	−	−	92 − 75	97 − 63	Test terminated at 1045 when effective temperature in control
0930	−	−	−	−	−	93.5 − 84	99 − 67	room reached 93.7°F
1000	−	−	−	−	−	94.5 − 87	99.5 − 73	Condition of men rapidly deteriorated
1030	−	−	−	−	−	95.9 − 90	99 − 76	Surface at 1115
1045	−	−	−	−	−	96.0 − 90	−	

*Temperature of air layer next to the skin over chest area (the personal climate)

†A special thermometer placed in the shoe at the instep.

‡"Hot" spaces are the engine rooms and maneuvering room. Although air is recirculated throughout the ship, the temperatures in control room and living spaces are lower since there are no hot surfaces which radiate heat.

one to two tons of refrigeration from the cold-storage food refrigerator, he could reduce humidity from an objectionable 90% (in tropical waters), to a comfortable 60%. By 1940, one could report the following results from an operation test [9]:

> "Ventilation and air conditioning of the submarine are adequate so that air conditions would not be a limiting factor in the length of wartime patrol which such a ship could make in tropical waters.
>
> "If one of the two air-conditioning units should fail, conditions closely approach a critical effective temperature and would be a limiting factor in the length of a submerged patrol which could be carried out in tropical waters.
>
> "If both of the 4-ton air-conditioning units should fail, the air conditions in the control and maneuvering rooms soon reach a critical effective temperature. Isolated endurance tests might be continued for periods of 11 or 12 hours, but it would be advisable to surface in about 2 hours in order to maintain personnel in condition for prolonged wartime patrol.
>
> "An unexpected finding was the negligible bacterial content of the cooled and recirculated air. Not a single case of upper respiratory tract infection occurred during nine days of submerged cruising. Apparently moisture removed by cooling coils carries with it dust and bacteria."

It was these efforts to provide remedial air conditions that comprised the most important single measure to render the physical condition of our personnel superior to those of the enemy. In this nuclear era a submerged cruise of 71 days duration in tropical, temperature, and arctic waters, and over distances covering the circumnavigation of the globe, is accomplished with less hardship than a three-day cruise in an S-boat 30 years ago. During a 60-day submerged cruise involving 116 men in a synthetic atmosphere, the air temperature was maintained at 73°F, humidity at about 50%. The average CO_2 concentration was 1.1%, CO 40 ppm, Freon 40 ppm, and aliphatic hydrocarbons 3 ppm. There were no aromatic hydrocarbons of the benzene series. The total aerosol concentration was less than 0.5 μg per liter [10].

The days lost by illness amounted to five for the entire crew; one man had a febrile respiratory infection. There were 17 afebrile "colds" during the early part of the cruise, leaving the crew immune to respiratory ailments for the remainder of the voyage. This is in accord with the general experience during and prior to World War II, when during the first 10 days of a cruise "colds" were prevalent, but during the succeeding 50 days or more the crew members were

Fig. 2. Transport berthing space [Courtesy Rear Admiral Walter Welham (USN)].

relatively free from respiratory infection. This immunity persisted until shore contacts were reestablished.

A bacteriologic laboratory set up in the torpedo compartment of one of the 1940-era submarines provided data to support the conclusion that the bacterial content of the cooled and recirculated air in these submarines was negligible. The condensate amounted to about five gallons per hour of which three-fifths was of human origin and the remaining two-fifths from the batteries. The batteries were frequently "overventilated" to reduce the danger of hydrogen explosion. Apart from the removal of bacteria from air by condensation of water vapor and aerosols, the oil film covering most of the surface of a submarine rendered the dust count negligible.

In some of the nuclear ships there has been a continuance of infections for the entire cruise. Preliminary tests have shown an increase in bacteria in recirculated air which was related to the time of submergence. The organisms identified were not respiratory pathogens [11].

Aerosols and Tobacco Smoke

In an exemplary analysis, Anderson and Ramskill [12] correlated the aerosol content of a nuclear submarine with the smoking of

tobacco.   Tobacco smoke is responsible for 75% of the aerosol
concentration.   The median diameter of the aerosols found was about
0.45μ (0.2 to 1.0μ), a size range characteristic of aged cigarette
smoke.   The chemical nature of the concentrated aerosol substance
was further identified as originating from cigarette smoke.   With
respect to gaseous compounds, 15 such substances appearing in
cigarette smoke have been identified in nuclear submarine atmospheres.
Both the aerosol and gaseous substances of cigarette origin affect
performance of men and equipment.   The electrostatic precipitators
are more than 95% effective in removing aerosols.   Nevertheless,
smoke from some 2000 cigarettes burned daily remains the prime
nuisance to contaminate the submarine atmosphere — a deleterious
nuisance to men and equipment of human origin and an inexcusable
one that could be interdicted by two words.

Mechanisms Required to Purify a Submarine Atmosphere [13]

The concentration of most of the contaminants in the submarine
atmosphere, whether in the form of solid or liquid aerosols or
present as gases, are maintained at a low level by the air purification
system consisting of scrubbers, catalytic combustion units, filters,
absorbers, cooling coils, and precipitators.   In addition, the concentration of some of the particulate matter is normally reduced
through agglomeration and settling, and by collision with surfaces.

CONTROL OF ATMOSPHERIC VARIABLES IN THE TREATMENT OF DISEASE

General Statement

During the past 25 years air conditioning has been extensively
applied as a valuable and at times indispensable adjunct in the therapy
of various diseases.   Some of the important applications which I
summarized at one time in a section of the "Guide of the Society of
Heating and Ventilating Engineers" [14] pertain to operating rooms,
nurseries for premature infants, maternity and delivery rooms,
children's wards, X-ray rooms, and wards for control of allergic
disorders.   During World War II the commissioning of hospital
ships featuring air conditioning of all wards, laboratories, and living
spaces constituted a notable achievement to improve patient therapy,
especially those afflicted with extensive burns.   At the present time
approximately 75% of all new hospitals are being air conditioned [15].

Importance of Air Conditions in the Treatment of Burns

At rest under euthermic conditions, about 40 cc of fluid is evapora-
ted per hour through intact skin as insensible water loss [16].   In burn
patients this fluid loss is tremendously augmented, and in hot environments fluid loss is measured in liters.   Under these conditions the

maintenance of the critical fluid and electrolyte balance of the body is jeopardized. Patients, for example, with second-degree burns (destructive of superficial layers of skin) have been shown to have a 20-fold increase in evaporative water loss compared with that from intact skin [17]. By means of deuterium oxide it has been shown that water vapor moves inward through intact skin above a relative humidity of 86% (the neutral relative humidity defined by Buettner) and that water is lost below this percentage. In addition to water vapor, all gases dissolved in the tissues of the body or in ambient air will diffuse outward or inward in response to gradient pressures.

Control of temperature and humidity is essential in the treatment of burned patients, and under tropical conditions, this measure is life-saving. Such control however has not been fully exploited in the "open air" treatment of burns, and surgical textbooks may contain no more than a scanty statement that the room should be without drafts and ideally at a temperature of 70 to 75°F. Reports are not available to support the routine value of oxygen in envelopes surrounding burned areas, the control of carbon dioxide, or use of bacteriostatic agents as aerosols in conjunction with ultraviolet light.

Air Conditioning of the Nursery

One of the notable milestones in the development of a remedial atmosphere was the painstaking investigation by Blackfan and Yaglou [18] of the importance of control of temperature and humidity in the early days of infant life. The provision of optimal air conditions reduced overall mortality by 75%. For the first time in medical therapy, the importance of the amount of moisture in the air was evident. The degree of humidity best suited for stabilization of body temperature of the infant was shown to be about 65%. A humidity of 30% was associated with instability of body temperature and other untoward effects leading to serious consequences.

The Ward

The data in Table IV were recorded on cardiac patients (for

Table IV. Mean Values of Blood Pressure and Pulse Rate of Cardiac Patients in a Naturally Ventilated and in an Air-Cooled Hospital Ward in New Orleans (Data from Burch and DePasquale [19])

Number of patients	Air conditions	Blood pressure	Pulse rate	Body temperature	Air temperature, dry-bulb	Relative humidity
88	Air-cooled	130/77	84	98.4	75	66
75	Natural ventilation	135/76	90	98.7	84	84

comparison with shipboard data in Table II) in New Orleans in a naturally ventilated ward and in one with air conditioning [19]. The lowering of pulse rate of six beats per minute appears to be small but it is appreciable in conserving the work energy of the heart over long periods and constitutes an essential factor in convalescence. The ability of the patients to rest and sleep without sweating is a prime advantage, as also shown in the Navy tests of the artificially cooled environment.

Ultraviolet Light Sterilization of the Operating Room

A noteworthy achievement pioneered by Dr. Deryl Hart at Duke University Medical Center is the continuous irradiation of the operating room and its occupants by ultraviolet light applied in effective concentrations [20]. The greatest percentage of contamination found in operating room air emanates from the surgical staff and results from their activities prior to and during surgical procedures [15]. These organisms may then be carried over the surgical field. Viewing an irradiated operating room through the overhead glass enclosure, one may appreciate the radical departure from usual surgical practice in controlling bacterial contamination when one observes the lack of covering needed for surgical instruments which are left in the open.

INHALATION THERAPY

In this rapidly expanding field well-trained technical personnel strive to improve apparatus and procedures for medical emergencies and routine therapy [21, 22, 23]. The value of oxygen inhalation is now so firmly established that facilities for its safe employment constitute an essential part of the hospital armamentarium. The early promise that carbon dioxide mixtures would be used routinely has not been realized. In high concentration carbon dioxide has the remarkable property of increasing blood flow through the brain. The use of helium mixtures is based on the decreased resistance of such mixtures approximately proportional to the square root of the density in their passage through the small lung recesses to alveoli. In the treatment of status asthmaticus, the use of helium—oxygen mixtures may be life-saving. The employment of gases will be discussed subsequently in connection with hyperbaric therapy.

The utilization of various medications as aerosols such as bronchodilators, wetting agents, antibiotics, and mucolytic agents is now accepted practice [22, 23]. It is possible to reproduce experimentally and for clinical purposes any parameter of the natural aerosol system. A noteworthy advance pioneered by Dautrebande [24] is the use of submicroscopic particles (less than $0.2\ m\mu$ in concentrations of 20 million particles per cc of air). The amount of drug given can be small enough to have no systemic effect yet its extremely fine size makes dispersion possible throughout the entire area of the lung.

Aerosol therapy has had application in the employment of vaso-dilators (epinephrine, isoproterenol). Antibiotics such as penicillin administered as aerosols present the hazard of sensitization reaction. There is no proof moreover of their entry into diseased alveoli of patients with pneumonia, and any improvement of these patients during inhalation therapy may be attributed to the absorption of the antibiotic into the blood stream.

Detergent and wetting agents have been introduced in the effort to loosen secretions. Such wetting and detergent agents as N-acetyl cysteine have been effective in the treatment of cystic fibrosis. The remarkable extent to which substances have been incorporated into aerosols is indicated by preparations from solutions containing ascorbic acid, sodium percarbonate, and copper sulphate. Seemingly the single-layered epithelium in the lung is treated as though it were skin.

Maintenance of Equipment and the Danger of Bacterial Contamination

Two matters require special emphasis, namely, proper care of equipment and the need for its careful sterilization [22, 23]. In an examination of nebulizers and respirators 85% of the machines produced aerosols with bacterial counts greater than those in surrounding air with Gram-negative (pseudomonas) predominant. All too often the organisms are penicillin-resistant and careful sterilization is required to avoid serious infection.

THE HYPERBARIC ENVIRONMENT – A GIANT REAPPEARS ON THE HORIZON

The quest of man for healing vapors has been no less intensive than the search for healing waters. Medical cabinets have been in use for more than 100 years. From the time of Robert Boyle and the introduction of the vacuum pump, the low-pressure chamber has been utilized for medical treatments. The advent of the high-pressure chamber in connection with caisson and diving operations of the last century almost immediately became the concern of physicians inter-ested either in the scientific or empiric practice of medicine. The remarkable engineering features embodied in Cunningham's chamber built in Cleveland in 1928, comprised 72 rooms housed in a steel sphere 50 ft in diameter. Engineering accomplishment at the time however excelled sound medical practice and the magnificent chamber was subsequently demolished for scrap iron.

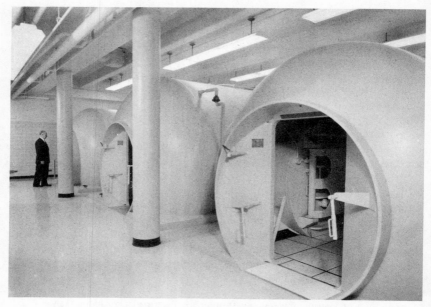

Fig. 3. Hyperbaric chambers, Lutheran General Hospital, Park Ridge, Illinois.

The renascence of hyperbaric therapy stems from the efforts of Dr. Boerema and his colleagues in Holland [25, 26]. They constructed a pressurized operating room and focused attention not only on the value of high oxygen pressures in modern surgery, but notably on the miraculous arrest of fulminating gas gangrene infection. The recovery brought about by hyperbaric oxygen in highly toxic, dying patients whose tissues are undergoing mass liquefaction by bacterial toxin is without precedent in medicine. The immediate result of Boerema's work was the construction of several hyperbaric medical centers and various pressurized facilities in connection with hospitals in the United States and elsewhere in order to evaluate critically the possible benefits accruing from a pressurized environment.

Treatment of Scuba Divers and Medical Emergencies

Without detracting from the importance of the hospital hyperbaric center, I am certain that one of the chief benefits resulting from application of pressures greater than atmospheric will be provision for on-the-spot treatment of victims of diving accidents and medical emergencies [27]. Millions of Americans, many of them adolescents, participate in sport diving and too frequently engage in hazardous dives with Scuba (Self-Contained Underwater Breathing Apparatus) gear. This equipment imposes the stringent requirement that a diver breathe through a mouthpiece without protection of a helmet. With the diver's head thus immersed in water, it follows that any mishap is attended by the probability of drowing or forced ascent to the surface without adequate decompression. Inadequate decompression results in bends and not infrequently induces widespread bubble formation in the blood stream to bring about asphyxia or paralytic injury to the spinal cord. A small, one-man portable chamber, pressurized with oxygen to no more than three atmospheres forestalls injury and possible death. Such a facility should be available to every diving group.

The same type of portable chamber can be made available for emergency oxygen therapy in cases of carbon monoxide asphyxia, coronary thrombosis, asphyxia of the newborn, and for vitalization of damaged tissues following crash injury and other accidents.

Pressure Effects Per Se

Compression therapy has been pointed out as life-saving in the treatment of moribund divers whose blood stream may be filled with gas bubbles. There are other benefits to be derived from pressure itself. The ventilization of lungs, sinuses, and middle ear spaces initially appeared to be one of the promising applications of hyperbaric therapy [28]. However, apart from the diagnostic value of chamber application of pressure for the detection of blockage of egress to

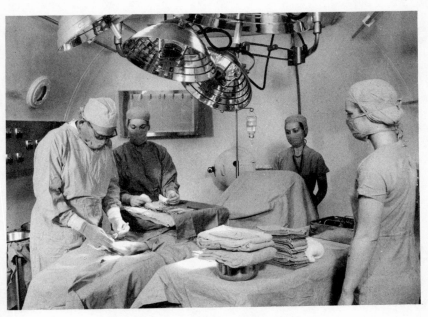

Fig. 4. Surgical team in hyperbaric chamber.

sinal and middle ear spaces, limited use has been made of this modality. Recently improvement of patients with chronic pulmonary disease (emphysema) has been reported following short exposures in compressed air at two atmospheres [29]. The increased partial pressure of oxygen in the air was insufficient to account for the benefits which in a manner as yet unexplained may persist for several weeks following treatment.

HYPERBARIC OXYGEN THERAPY

Turning now to the main-line effort in the hyperbaric field, it may be appropriate to comment briefly rather than present a comprehensive review of some applications to indicate the scope of the possibilities in this type of therapy. The rationale underlying the administration of oxygen at elevated pressure relates to benefits accruing from increased oxygen pressure in cells. Although oxygen can be inhaled for short periods at least at pressures sufficient to relieve blood cells of their function in oxygen transport, it has not been possible to measure oxygen pressure at the cellular level. The toxicity of oxygen moreover restricts inhalation usually to pressures of three atmospheres or less, and for sustained inhalation, to time periods of about two to three hours. Nevertheless, although these conditions are restrictive, it is possible to perform surgical operations and to treat asphyxial conditions. If the body is cooled to lower temperatures, the oxygen requirements are reduced and the toxicity of oxygen is decreased. In deep hypothermia, oxygen is not released adequately from red blood cells and hence the pressure of oxygen in physical solution brought about by the hyperbaric atmosphere is paramount. Under conditions of deep hypothermia combined with hyperbaric oxygen, the heart and brain have recovered their function following deprivation of their blood supply for periods of more than 30 minutes. Although histologic damage may be found in such blood deprived tissue, it is apparent that the synergistic combination of hypothermia and hyperbaric oxygen may be routinely life-saving.

Dr. Claude Hitchcock has opened the way to preservation and safeguarding of tissue *in vivo* and postmortem by means of hyperbaric oxygen and hypothermia. If such tissue can be perfused, it may be preserved for subsequent implantation months and even years later. An old individual in generations to come may have the benefit of implantation of his own tissue which was removed during adolescence and preserved in the hypothermic, hyperbaric atmosphere.

Asphyxial Conditions

It was J. B. S. Haldane [30] who showed many years ago that rats would survive in oxygen at three atmospheres even though their hemoglobin was saturated with carbon monoxide. The value of this

classical application of hyperbaric oxygen has been repeatedly confirmed, and patients severely poisoned from coal gas have recovered
in the hyperbaric chamber.

In the treatment of early or advanced asphyxia of the newborn,
Hutchinson and Kerr [31] found that it was possible by means of a
small chamber to raise oxygen pressure to desired levels and to
bring about recovery that was frequently dramatic. "This method
has considerable advantage over tracheal intubation in that the
apparatus can be operated by the physician after only a very short
course of instruction. It can be used quickly, and in our more recent
cases, compression has been started whenever effective respiratory
efforts have not been established within four minutes of birth. On
purely physiologic grounds, the treatment of anoxia by hyperbaric
oxygenation would seem to be logical. The development of efficient
apparatus has made its therapeutic implementation practicable"[31].

Anaerobic Infections — Gas Gangrene

The striking effects of hyperbaric oxygen in terminating the ravages
of gas gangrene have been repeatedly confirmed. The clostridial
organisms are not killed by oxygen, but oxygen inhibits the production
of the destructive tissue-liquefying toxin.

The inhibitory effect of oxygen on anaerobic organisms has long
been known but only recently by means of the hyperbaric chamber
have favorable results been obtained in infected patients. The
bacteriostatic action of oxygen on anaerobic organisms holds promise
for the control of putrefactive bacteria in the respiratory tract, in
treatment of skin infections where many atmospheres of oxygen may
be employed, and in the intestinal tract itself where high concentrations
of oxygen can be introduced by duodenal intubation or rectal administration.

Hyperbaric Surgery

The concept that life is possible without red blood cells even for
short periods of operative surgery has not as yet been unequivocally
established in the hyperbaric chamber. Nevertheless, the number of
red blood cells normally required for extracorporeal circulation may
be greatly reduced. Surgery to correct congenital abnormalities of
the heart in children has been carried out [32] in the same chamber at
the Harvard School of Public Health where investigations some 30
years ago were carried out in exposure of healthy men to high oxygen
pressures. Despite the restricted space and the difficulty of converting
a test chamber into an operating room, the results to date justify
the employment of oxygen at high pressures to correct cardiac defects
in cyanotic infants and children who are critically ill.

Restorative surgery in the hyperbaric atmosphere has saved

limbs of patients who have had crushing injuries and those who have had occulsive vascular disease. Techniques for supplying oxygen to tissues remain to be perfected and consideration must be accorded the fact that although hyperbaric oxygen is a prime modality, it is but one of a number of critical therapeutic procedures required during the course of intricate, reparative surgical procedures.

Hyperbaric Oxygen and Radiotherapy

It has been established that a cell deprived of oxygen is nearly three times less sensitive to damage by X rays. Malignant tumor cells which are known to exist in a hypoxic environment may therefore be protected against the lethal effect of radiation administered in doses which do not destroy normal tissue. In order to raise the oxygen pressure in tumor cells, patients with advanced carcinoma have been irradiated through the walls of plastic chambers filled with oxygen at pressures of two to three atmospheres. The results of this type of therapy carried out by Dr. Churchill-Davidson [33] have been encouraging, notably in sterilization of primary tumors of the head, neck, and mouth. The number of persons thus far who have survived for more than four years from initiation of therapy has justified and more than compensated for the difficulties involved in prosecuting this type of therapy.

HYPERBARIC THERAPY WITH GASES OTHER THAN OXYGEN

In the undersea environment it is being demonstrated that men can live for days at depths of 400 ft or more (13 atmospheres or higher) in helium-oxygen atmospheres. The feasibility of utilizing such pressures is a challenge to conduct research with gases other than oxygen to affect cellular function favorably. Gases can penetrate cellular membranes which are impermeable to solutes. Thus far the gases studied apart from oxygen have been physiologically inert. This inertness did not prevent a narcotic action which we described many years ago in connection with the effect of nitrogen in air at high pressures [34]. Basic mechanisms of anesthesia have subsequently been revealed by observations of the hyperbaric effects of inert gases. With respect to the narcosis, more widely known by Cousteau's characterization, "L'ivresse des grands profundeurs" (freely translated as "rapture of the depths"), one obtains a revealing insight into human behavior under conditions of a mild and harmless stress. A systematic evaluation of function and personality traits of persons subjected to the stress of the hyperbaric environment has only begun.

DIFFUSION OF GASES THROUGH SKIN

Mention was made of the movement of water vapor inward through the skin when the relative humidity was higher than 86%. The significance of diffusion of gases through the skin became apparent in experiments with helium. If an individual was placed in a rubber bag with his head protruding and the bag filled with helium, then helium diffused through the skin into subcutaneous blood capillaries and thence by transport to the lungs. The amount of helium eliminated from the lungs could be determined. On the other hand, if a helium—oxygen mixture were inhaled, the helium transported from the lungs to the skin and thence by diffusion through the skin could be collected and measured in the rubber bag surrounding the body to the neck level.

Although the diffusion of gases through intact skin is appreciable, it is increased manyfold if the skin is incised or presumably if its integrity is destroyed by thermal burn. Under conditions of injury to the skin it should be possible to supply oxygen in large quantities to body tissues percutaneously at pressures of 13 atmospheres or greater. If in addition oxygen is introduced into the gastrointestinal tract, pulmonary function may not be required for oxygenation of red blood cells.

CONCLUSION

1. The cursory comments in the preceding paragraphs pertain to the wide scope of hyperbaric oxygen therapy and to the many competent clinical investigators engaged in evaluation of this type of therapy. It is too early to project the ultimate gains to be derived from the hyperbaric environment. The minimal five-year period generally required for evaluation of drug, surgical, or mechanical therapy has only been extended to several applications of hyperbaric oxygen pressure. At this time the field has a potential limited only by the imagination, ingenuity, and training of the investigators concerned. One may agree with Henshaw [35] who inaugurated hyperbaric therapy in 1664 that there is apparent justification for current enthusiasm, but to gain a worthy place in the therapeutic armamentarium the test of time will require that this modality meet at least three criteria: it must be uniquely beneficial and have clear-cut advantages over less formidable forms of treatment; it must be fully practical in its use in at least a few important conditions; and in consequence, it must be applicable to a reasonable number of patients.

2. The elaborate and expensive units required to "condition" contaminated air emphasizes the need to control contaminants at their source. An experienced surgeon once said that the way to keep one's hands clean was not to contaminate them. Experience in hospitals

has proved how difficult it is even with every method of cleaning and sterilization employed to get rid of highly pathogenic organisms. As in the submarine, so in a metropolitan area, the various units available within the economy of the user can only partially clean air from a heavily contaminated source.

The persisting respiratory infection reported in some nuclear submarines and the inability to "sterilize" submarine air despite the operation of filters, burners, precipitators, scrubbers, and air conditioning make mandatory a careful study of droplet infection.

In regard to the individual there is a rationale in using a simple face mask to protect others against droplet infection. It may be necessary in civilian life to provide unit ventilation to individual cubicles constructed of transparent material to protect key personnel in daily contact with many persons.

ACKNOWLEDGMENT

As a former submarine medical officer, I am deeply indebted to Captain T. H. Urdahl, USNR (Ret), at one time in charge of the Air-Conditioning Section of the Bureau of Ships. Captain Urdahl's insight, judgment, and engineering knowledge served to disrupt the chronic *status quo* attitude to bring about the striking improvement in personnel welfare and efficiency alluded to in this paper.

REFERENCES

1. Goldsmith, J. R. "Epidemiologic studies of obstructive ventilatory disease of the lung," Amer. Rev. Resp. Diseases Vol. 82, p. 485, Oct., 1960.
2. Tromp, S. W. (Ed.) Medical Biometeorology. American Elsevier, New York, 1963.
3. Licht, S. (Ed.) Medical Climatology. Waverly Press, Baltimore, 1964.
4. Mills, C. A. "Physical environment and effective manpower," Military Surgeon, Vol. 80, p. 331, May, 1937.
5. Hollander, J. L., and Erdman, W. J. II "The controlled-climate chamber," Medical Climatology, S. Licht, Ed., Waverly Press, Baltimore, 1964, p. 702.
6. Tromp, S. W. "The application of simulated high altitude climate in a low pressure climatic chamber to asthmatic and bronchitic patients," Vol. VIII, Biometeorological Research Center, Leiden, 1964. (See also ref. 2: Medical Biometeorology, S. W. Tromp, (Ed.).
7. Yanda, R. L., and Herschensohn, H. L. "Changes in lung volumes of emphysema patients upon short exposures to simulated altitude of 18,000 feet," Aerospace Med. Vol. 35, p. 1201, Dec., 1964.
8. Submarine (The), Manual Prepared by the Submarine School, New London Submarine Base, Groton, Conn. NAVPERS 16160-B. Washington, D. C., Govt. Print. Office, p. 89 in current edition not dated.
9. Behnke, A. R. Unpublished data from submarine operations, 1940.
10. Ebersole, J. H. "Submarine medicine on the Nautilus and Seawolf," AMA Arch. Int. Med. Vol. 18, p. 200, 1958. "New dimensions in submarine medicine," New Eng. J. Med. Vol. 262, p. 599, Mar., 1960.
11. Anderson, W. L. "Aerosols in nuclear submarines," Chap. 18 (Annual Progress Rpt., NRL Rpt. 5630), The Present Status of Chemical Research in Atmosphere Purification and Control on Nuclear-Powered Submarines. V. R. Piatt and E. A. Ramskill (Eds.), 14 July 1961, U. S. Naval Research Laboratory, Washington, D. C.

12. Anderson, W. L., and Ramskill, E. A. "Aerosols in nuclear submarines," Chap. 20, (NRL Rpt. 5465), 20 April 1960, Washington, D.C.

13. White, J.C. "Atmospheric control in the true submarine," (Report of NRL Progress 1-16). U.S. Naval Research Laboratory, Washington, D.C., 1958.

14. Behnke, A. R. "Air conditioning in the prevention and treatment of disease," Chap. 13, Heating, Ventilating, Air-Conditioning Guide. Amer. Soc. Heating and Ventilating Engineers, New York, N.Y., 1946.

15. Gaulin, R. P. "How to keep infection out of the air," Modern Hospital, Vol. 100, p. 93, 1963.

16. Burch, G.E., and De Pasquale, N. Hot Climates, Man and His Heart. (Chap. 4, p. 22), Charles C. Thomas, Springfield, Ill., 1962.

17. Neely, W. A., Turner, M. D., Smith, J. D., and Williams, J. Surgical Forum, Vol. 12, p. 13, 1961.

18. Blackfan, K. D., and Yaglou, C. P. "Premature infant: Study of effects of atmospheric conditions on growth and development," Amer. J. Dis. Child. Vol. 46, p. 1175, Nov., 1933.

19. Burch, G. E., and De Pasquale, N. "Influence of air conditioning on hospitalized patients," J. Amer. Med. Assn. Vol. 170, p. 160, May, 1959.

20. Hart, D. "Sterilization of the air in the operating room with bactericidal radiation," Arch. Surgery Vol. 41, p. 334, Aug., 1940.

21. "Treatment by inhalation," The Lancet, Vol. 2, p. 1205, Dec., 1962.

22. Lovejoy, F.W. Jr., and Morrow, P.E. "Aerosols, bronchodilators, and mucolytic agents," Anesthesiology Vol. 23, p. 460, Jul-Aug., 1962.

23. Segal, M.S., Traverse, N., and Dulfano, M.J. "Inhalational therapy for chronic lung disease," Anesthesiology Vol. 23, p. 513, 1962.

24. Dautrebande, L. Microaerosols. Academic Press, London, 1962.

25. Boerema, I. "An operating room with high atmospheric pressure," Surgery Vol. 49, p. 291, Mar., 1961.

26. Brummelkamp. W. H., Hogendijh, J., and Boerema, I. "Treatment of anaerobic infections (Clostridial Myositis) by drenching the tissues with oxygen under high atmospheric pressure," Surgery Vol. 49, p. 229, Mar., 1961.

27. Behnke, A. R. "Problems in the treatment of decompression sickness (and traumatic air embolism)," Ann. N. Y. Acad. Sci. Vol. 117, Art. 2, p. 843, 1965.

28. Behnke, A. R. "High atmospheric pressures: Physiologic effects of increased and decreased pressure. Application of these findings to clinical medicine," Ann. Internal Med. Vol. 13, p. 2217, 1940.

29. Yanda, R. L., Motley, H. L., and Smart, R. H. "The effects of pressure upon lung volumes of pulmonary emphysema patients and upon normal individuals," First International Congress on Clinical Application of Hyperbaric Oxygen, The Netherlands, University of Amsterdam, 1963.

30. Haldane, J. B.S. (1927) Cited by J.S. Haldane and J.G. Priestley, Respiration. Oxford, 1935, p. 238.

31. Hutchison, J.H., and Kerr, M.M. "Treatment of asphyxia neonatorum by hyperbaric oxygenation," Ann. N. Y. Acad. Sci. Vol. 117, Art. 2, p. 706, 1965.

32. Bernhard, W. F., and Tank, E. S. "Effect of oxygen inhalation at 3.0 and 3.6 atmospheres absolute upon infants with cyanotic heart disease," Surgery Vol. 54, p. 203, 1963.

33. Churchill-Davidson, I. "The small patient chamber, radiotherapy," Ann. N.Y. Acad. Sci. Vol. 117, Art. 2, p. 875, 1965.

34. Behnke, A. R., Thomson, R. M., and Motley, E. P. "Psychologic effects from breathing air at increased pressures," Amer. J. Physiol. Vol. 112, p. 554, 1934.

35. "Hyperbaric oxygen: Potentialities and problems," (Quotation from Henshaw [1664], Abstracted from Simpson, A. Compressed Air as a Therapeutic Agent. Sutherland and Knox, Edinburgh, 1867). Report of the Ad Hoc Committee on Hyperbaric Oxygenation, Nat. Acad. Sci., Nat. Res. Council., Washington 25, D.C., 1963.

The Environmental Spectrum—Today

Moderator: Harold B. Gotaas, *Dean, The Technological Institute, Northwestern University, Evanston, Illinois*

Heat: Warren Viessman, *Chemical Engineering Consultant, Air Force Systems Command, Washington, D.C.*

Acoustical: Edward R. Hermann, *Associate Professor of Environmental Health Engineering, Northwestern University, Evanston, Illinois*

Odor: Richard L. Kuehner, *Manager, Environmental Sciences, Borg-Warner Research Center, Des Plaines, Illnois*

Sensory: Max V. Mathews, *Director of Behavioral Research, Bell Telephone Laboratories, Murray Hill, New Jersey*

Bacteria: James G. Shaffer, Sc.D., *Director of Microbiology and Hospital Epidemiology, Lutheran General Hospital, Park Ridge, Illinois*

Man and His Thermal Environment

Warren Viessman

Directorate of Civil Engineering
Air Force Systems Command
Washington, D.C.

INTRODUCTION

The control of the thermal environment within limits, either by the addition or subtraction of heat, is conducive to the health and comfort of man. It reacts with and complements his own heat production system by heat interchange, and has a definite bearing on his ability and efficiency in performing assigned tasks. The interaction between body and space heat will be discussed and tolerances given for comfort and survival.

BODY HEAT

By biological processes, heat is generated in man and regulated to maintain a body temperature around 98.6°F. The total heat output consists of sensible heat and latent heat. Sensible heat is the heat associated with a change in temperature. Heat energy flows from a body at high temperature to a surrounding environment at lower temperature. In a room at 70°F, the skin temperature of a person would be about 92°F, and the clothing surface temperature 84°F. The body loses heat under these conditions, to produce a comfortable environment. Latent heat is the heat associated with a change of state, such as a change from a solid to a liquid, or a liquid to a gas. The unit of heat energy commonly used is the British thermal unit (Btu), and is defined as the approximate amount of heat required to raise the temperature of one pound of water from 59 to 60°F.

Heat is given up by man by three processes: (1) vaporization from the skin and lungs, (2) radiation from the skin, and (3) convection. For a person at rest, these processes constitute approximately 25, 50,

77

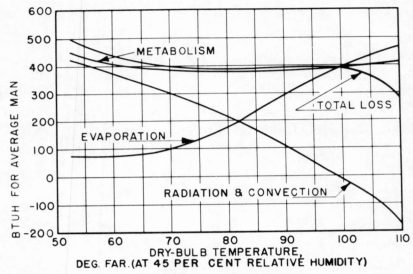

Fig. 1. Heat loss for average man at rest.

and 25% respectively of the total heat loss, depending on environmental conditions. At 70°F room temperature for a lightly clothed average man at rest, these would amount to about 100, 200, and 100 Btu/hr, or a total of 400 Btu/hr. This could increase to 2400 Btu/hr or more for the hardest sustained work. Loss by conduction from clothed persons is usually negligible. Lowering the temperature of the environment increases the body heat loss by convection and radiation. Lowering the humidity increases the latent heat loss. If the ambient temperature is raised until it reaches that of the body surface, convection and radiation become nil. When the ambient temperature is increased to a higher value than the body temperature, heat is given to the body by the environment.

In order to maintain normal body temperature and heat balance as the body is heated, the surface blood vessels expand, the output of heat increases, and there is an increase in blood flow to the skin. This process carries heat from the internal parts to the surface of the body, where it is lost by convection and radiation. At high ambient, above skin temperatures, the perspiration rate increases and heat in the form of water vapor is given up.

On a hot, dry day, man may produce a quart of sweat in an hour. When the loss in body weight approaches 5%, heat stress occurs, physical deterioration results, perception is distorted, and judgment fails. At 12% the metabolic heat can no longer be dissipated through the

skin and respiratory tract.   The body temperature then rises to a
fatal level.

Below 75°F for normally clothed people, and below 85°F for persons
lightly dressed, evaporative losses are minor.   The radiation and
convective losses increase with cooler environment.   In order to
maintain and regulate normal body temperatures at lower ambient
temperatures, there is a constriction of the blood vessels next to
the skin to keep the warm blood away from the surface of the body.
This results in cooling of the body surface and therefore reduces the
amount of heat lost to the environment.   In order to increase the
metabolic heat production, shivering occurs.   This, however, is never
sufficient to replace the heat lost.   The loss can be further counter-
acted by an increase in the insulating effect of clothing, or body
temperature can be increased by increase in metabolic heat through
exercise or work.   When thermal balance cannot be maintained, cold
stress occurs.

HEAT BALANCE

Body temperature depends on a balance between heat production
and heat loss.   Heat resulting from oxidation of food in the body
(metabolism) maintains the body temperature above the surrounding
air in a relatively cool environment.   Under these conditions, heat
lost from the body by radiation, convection, and evaporation equals the
metabolic heat.   When normal balance is not maintained, heat is either
stored in or taken away from the body.   Within limits and depending

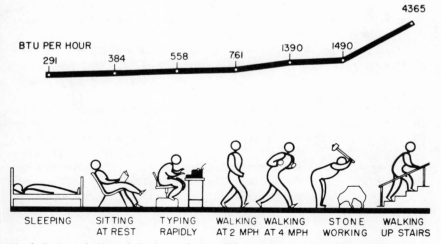

Fig. 2. Heat production of the human body in Btu per hour according to various activities.

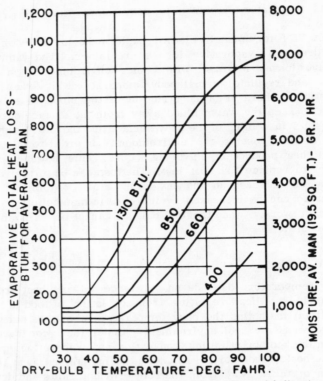

Fig. 3. Evaporative heat and moisture loss for various metabolic rates.

on activity, the body temperature is maintained within a normal range of 97 to 99°F.

The equation for the heat balance between the body and its environment can be written as follows:

$$M = \pm S + E \pm R \pm C$$

where M is the body heat, ranging from 250 Btu/hr for sleeping to 2400 Btu/hr or more for the hardest sustained work (a normal value for persons sitting quietly is 400 Btu/hr), S is the heat loss or gain in body heat storage, E is the evaporative heat loss, R is the radiation loss or gain, and C is the convection loss or gain. Normal conditions will prevail when this equation balances by thermal interchange of body heat with the environment and body temperature is maintained.

COMFORT TOLERANCES

A comfortable environment depends on three conditions: temperature, humidity, and air motion. An empirical index combining these

sensory effects into a single value has been developed and is documented by the American Society of Heating, Refrigerating, and Air-
Conditioning Engineers (ASHRAE).   It is known as the effective
temperature (ET).   Combinations of temperature, humidity, and air
motion that produce the same approximate feeling of warmth are
assigned the same effective temperature value. Figure 4 shows the
relationship existing between wet- and dry-bulb temperatures, relative
humidity, and effective temperatures for still air (air motion less than
20 ft/min).   The revised ASHRAE Comfort Chart (1963 Guide and
Data Book, Chapter 8, Figure 12, page 116) shows that about 80% of
the people are generally comfortable at effective temperatures between
65 and 70°F in winter and 68 to 73°F in summer. At these effective
temperature conditions, normal interchange or heat between a body
and its environment occurs, and the body temperature is maintained
within the 97 to 99°F range.

COLD STRESS

Human performance in extreme cold depends upon maintaining a

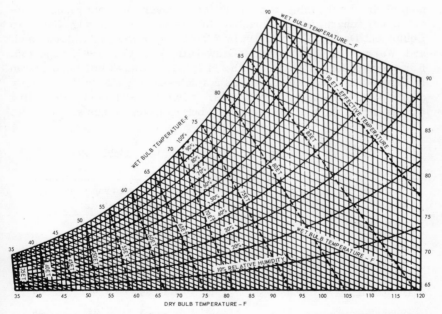

Fig. 4. Effective temperature (ET) psychrometric chart (Based on ASHRAE Psychrometric
Chart No. 1, 1963).   90 ET, upper limit for continuous exposure at light activity; 85 ET,
upper limit for moderately hard work; 80 ET, upper limit for heavy work; 78 ET, threshold
of sweating.   Applicable where radiation is negligible. Requires minus correction for
humidities at low temperatures and plus correction for humidities at high temperatures.

thermal balance between the subject and his environment. The ability to do so depends on the extent and duration of cold exposure, metabolic heat production, and heat conservation as affected by insulation or clothing. In cold environments with inadequate clothing, extreme discomfort and shivering occur, but the effects are not serious for short exposures. For prolonged cases of cold stress, freezing of extremities may first occur, followed by possible fatal action.

Insulation of clothing is conveniently expressed in terms of "clo" units. One clo, by definition, is the amount of insulation required to maintain a sitting—resting subject in comfort in a room at 70°F and 50% relative humidity. One clo is approximately the insulating value of a person's everyday lightweight business suit for mild climates. An insulating value of four clo represents arctic clothing. Clothing having insulating values in excess of four clo is impractical for active persons, due to bulk. Table I shows the clothing required for equilibrium during prolonged performance in outdoor environments.

No one index such as effective temperature is available to express all of the factors involved in cold exposure. A wind chill chart expressing the severity of the total cooling power of the environment was developed by Falkowski and Hastings in 1958 and has been found helpful in determining the effect of exposure to cold winds. This empirical chart is documented in Human Engineering Guide to Equipment Design, Chapter 10, by Albert Damon, Ross A. McFarland, and Warren H. Teichner, McGraw-Hill Book Co., 1963, page 435. For most practical purposes the simplified Wind—Chill Index shown in Figure 5 is adequate for use as a single-value guide to the severity of temperature—wind combinations. It is not based on human cooling and is probably not too accurate but is nevertheless serviceable.

Table I. Clothing Requirements for Cold Zone Personnel

Body clothing type	Clothing insulation, clo	Minimum enviromental temperatures, °F, for various exposure times*			
		Indefinite	5 hr	2 hr	1 hr
Light	0.8	70	65	55	45
Intermediate	2.3	45	35	20	0
Heavy	3.8	20	5	-15	-45

*Based on moderate activity, adequate hand, foot, and face protection at the lower temperatures and moderate winds.

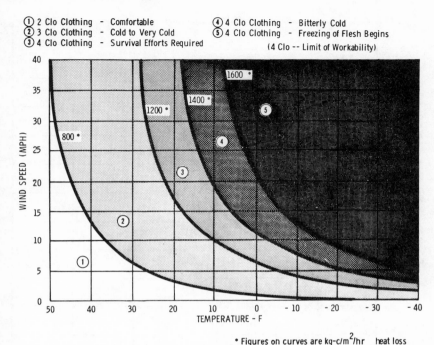

① 2 Clo Clothing - Comfortable   ④ 4 Clo Clothing - Bitterly Cold
② 3 Clo Clothing - Cold to Very Cold   ⑤ 4 Clo Clothing - Freezing of Flesh Begins
③ 4 Clo Clothing - Survival Efforts Required   (4 Clo -- Limit of Workability)

* Figures on curves are kg-c/m^2/hr   heat loss

Fig. 5. Wind chill index.

Man may become habituated to cold, i.e., learn how to survive during cold periods. He may also become acclimatized, i.e., physiological changes may be produced so as to make him more tolerant to cold. Available meager information indicates that man habituates to cold rather than acclimatizes to cold.

HEAT STRESS

The biological response to hot environments has been discussed under Body Heat. When men go to hot desert and jungle areas some adaptation to the climate takes place. If work is gradually increased daily in the hot environment, and if the men get adequate water, salt, and sleep, acclimatization may be complete in seven to 14 days. The acclimatized man works with a lower heat rate, lower skin and rectal temperature, and more stable blood pressure than one not acclimatized. Acclimatization is maintained at a high level for about two weeks after removal from the hot environment. Beyond this period it is necessary to be re-exposed to the environment at two-week intervals in order not to loose the acclimatization.

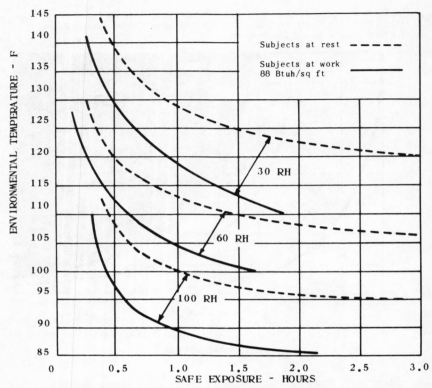

Fig. 6. Safe exposures for high temperatures at various humidities (Reproduced from 1963 ASHRAE Guide and Data Book by permission).

In fallout or protective shelters an effective temperature of 85°F is considered the maximum heat exposure for a 14-day period for the average occupant. In some industries it is not feasible, due to process requirements, to maintain optimum conditions of comfort. The problem usually resolves itself into one of alleviating conditions by one of the following means: (1) insulate the source of heat, (2) reduce temperature by bringing in cool air, (3) reduce humidity by venting steam or moisture to the outside, (4) strip down to a minimum amount of clothing, (5) limit hours of exposure, and (6) acclimatization of men to heat. All these methods assist in maintaining efficiency without impairment of health. Figure 6, reproduced from the ASHRAE Guide and Data Book, shows the safe periods of exposure for extreme environmental temperatures at several humidities.

The thermal load to which man is subjected is dependent on the interrelation between the metabolic heat rate and the environmental

heat as expressed in the above heat balance equation. If thermoregulation fails and the body heat storage increases, a person can
become incapacitated. The disability in the order of severity might
take the form of heat cramp, heat exhaustion, or heat stroke. The
stress imposed is mainly the result of the body's inability to achieve
a sufficiently high rate of evaporation by perspiration which begins
at an effective temperature of 78°F. A number of indices have been
developed to assess heat stress. While these are very helpful, no
single index seems completely reliable. Among the indices in use
are the effective temperature index, the Robinson index of physiological
effect, the four-hour sweat rate index, and the Belding—Hatch heat
stress index. These are documented in the ASHRAE Guide and Data
Book, Human Engineering Guide to Equipment Design, Environmental
Engineering for Fallout Shelters, Civil Defense, TR-23, by Prof.
R. G. Nevins, page 74, and elsewhere. They are recommended
reference material on the subject.

SUMMARY

For comfort, health, and efficient performance, the metabolic
heat production of man must be in balance with his thermal environment. Heat must be removed at the rate it is generated. If the heat
potential of the environment is below that of the body, heat is given
up to the environment by convection, radiation, and evaporation. If
the potential is such that the metabolic heat rate cannot maintain a
body temperature of 97 to 99°F, the person suffers cold stress. Where
it is necessary to be exposed to such cold conditions, clothing and
wind protection are required to alleviate the symptoms. While man
habituates to some extent to cold, prolonged exposure may be fatal.
The lowest survival temperature varies with age, activity, wind
velocity, clothing, sun or shade, altitude, and duration.

Heat stress results when the heat potential of the environment is
above that of the body and the metabolic process cannot maintain
normal temperature. In this case the environment heats the body and
the body temperature rises. Protection is afforded by insulation of
the person from the heat source, ventilation or air conditioning,
removal of humidity, use of a minimum amount of clothing to increase
radiation and convection, limit of duration, and acclimatization.

Acoustical Aspects of the Environmental Spectrum*

Edward R. Hermann

Civil Engineering Department
Environmental Health Engineering
Northwestern University
Evanston, Illinois

INTRODUCTION

Acoustics pertains to the effect of sound on hearing. The environmental spectrum with which we are concerned includes any space which may be successfully occupied by man. For the sake of time and specificity, this discussion will be limited to the more mundane aspects of excessive noise exposure. That is, those aspects measurably affecting the most people in their ordinary occupational, avocational, and recreational pursuits. The psycho-acoustical effects of noise will not be discussed, but rather the noise intensities loud enough to produce temporary or permanent threshold shifts in human hearing acuity will be considered. Temporary threshold shifts (TTS) and permanent threshold shifts (PTS) may be illustrated by the idealized scheme shown in Figure 1. Here we see tne hypothetical build-up of a cumulative effect due to repeated insults.

AUDIOMETRY

Although many hi-fi enthusiasts suffer from the delusion that their own hearing is 20/20 (a perfectly flat decibel response at all frequencies from 20 cps to 20 kcps), the sad facts are: (1) Man's

*Presented before the Northwestern University Conference: Interactions of Man and His Environment, January 28-29, 1965, Chicago, Illinois.

Preparation of this paper was supported in part by PHS Grants Nos. PHT 2-21 and OH-00178 from the Bureau of State Services, Divisions of Community Health Practice and of Occupational Health, respectively, U.S. Public Health Service.

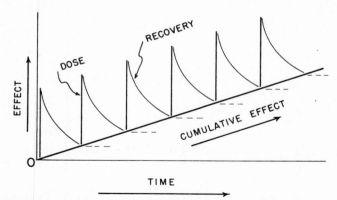

Figure 1. Idealized scheme illustrating build-up of a cumulative effect from repeated insults to a physiological system.

hearing above 10,000 cps is frequently poor to nonexistent; (2) the decibel—frequency response curve for the average of persons possessing unimpaired normal hearing is quite distorted relative to a single reference sound pressure level; and (3) man (especially the male of the species) progressively loses high-frequency hearing acuity starting at about 25 years of age. (Marriage!)

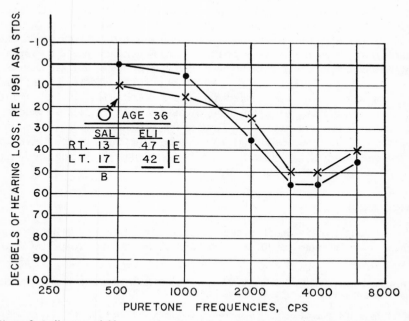

Figure 2. Audiogram of 36-year-old male subject. Results obtained by manual audiometry.

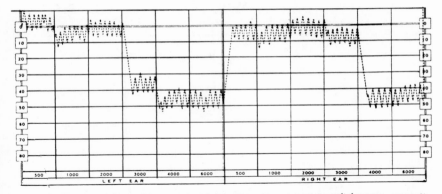

Figure 3. Audiogram obtained by use of an automatic audiometer of the Békésy type, as modified by Rudmose.   Subject: 40-year-old tube cleaner, 12 years on job; claims he wears ear protection.

On the other hand it must be realized that the human ear is an extremely sensitive yet rugged and durable sound-sensing device that can detect threshold vibratory pressure levels as low as 0.2 millidyne per square centimeter (3 billionths of one pound per square inch). It can operate for more than a quarter of a century over sound-pressure variations extending to more than 20,000 times as great as its threshold of detection with no appreciable deterioration. The human ear can even withstand sound-pressure variations more than a million times its threshold of detection without permanent damage if the exposure time is limited.

As with the other human sense organs, the ear does not possess absolute calibration. Fortunately, engineering development of precise low-cost electronic equipment which can be calibrated may be coupled with the human ear to provide noise-control criteria based upon changes in hearing acuity.   Pure-tone audiometry has been most widely used for this purpose; audiograms obtained by manual and automatic testing procedures are shown in Figures 2 and 3, respectively.   The fundamental sounds of major importance produced by the human voice range in pitch from 50 to 2500 cps.   It has been found that audiometric testing at the three pure-tone frequencies of 500, 1000, and 2000 cps provides an appraisal of the acuity of human hearing in the speech-important frequencies that is adequate for most purposes. The speech average loss (SAL) method for grading hearing acuity in conversational tones is based upon hearing threshold determinations at these three frequencies [1].   Table I relates audiometric findings with a grading scale and briefly indicates corresponding hearing characteristics.   This technique and grading scale was developed by

Table I. Speech Average Loss, Conversational Audiometry
(NRC Committee on Hearing)

Grade	SAL, db	Class name	Characteristics
A	< 16 Worse ear	Normal	Both ears within normal limits No difficulty with faint speech
B	16-30 Either ear	Near normal	Has difficulty only with faint speech
C	31-45 Better ear	Mild impairment	Has difficulty with normal speech but not loud speech
D	46-60 Better ear	Serious impairment	Has difficulty even with loud speech
E	61-90 Better ear	Severe impairment	Can hear only amplified speech
F	>90 Better ear	Profound impairment	Cannot understand even amplified speech
G	Total deafness in both ears		Cannot hear sound

Note: A person is graded one class lower than indicated by the above scale if, with an average loss of 10 db or more, his range of hearing threshold in the three speech frequencies is 25 db or more in both ears considered separately.

Table II. Early Loss Index, 4000 cps Audiometry

Age-specific presbycusis, db			ELI scale		
Age	Women	Men	Grade	Exceeds ASPV by:	Remarks
25	0	0	A	< 8 db	Normal-excellent
30	2	3	B	8-14	Normal-good
35	3	7			
40	5	11	C	15-22	Normal-within expected range
45	8	15			
50	12	20	D	23-29	Suspect noise-induced loss (NIL)
55	15	26			
60	17	32	E	30 or more	Strong indication of NIL
65	18	38			

the Committee on Hearing and Bioacoustics of the National Research Council. Although there is growing acceptance in the United States that hearing impairment should be based upon hearing threshold levels at the 500, 1000, and 2000 cps frequencies, it must be realized that initially human hearing is most sensitive in the region of 4000 cps, and also is most susceptible to damage in this region. On this premise, a quantitative method for measuring hearing losses among the noise-exposed has been developed which enables the prevention of significant losses in the speech-important frequencies. Without delving into the mathematics that enabled development of this early loss index (ELI) for noise-induced hearing loss, suffice it to note that if the age-specific presbycusis value (ASPV) is subtracted from the measured threshold of hearing at 4000 cps, a number is left that is statistically indicative of noise-induced hearing loss. Analysis of some 150,000 items of data obtained by audiometric testing of 5000 oil refinery employees in 14 different occupational classifications provided validation of the investigator's hypotheses [2]. A grading scale for the early loss index is shown in Table II.

Summation diagrams for comparing ELI and SAL class distributions are shown in Figures 4, 5, and 6. Due to the fact that high-frequency hearing loss as characterized by the 4000 cps notch precedes noise-induced hearing losses in the speech-important frequencies, the human ear itself may be used as an integrating dosimeter of noise exposure with very little sacrifice of overall hearing ability. The disadvantage of such a practice is obvious; however, the advantages presently far outweigh the undesirable aspects. Such variables as susceptibility to noise-induced loss, diligence of individuals in using available ear protection, or limiting time of exposure are obviated by using the audiometric approach rather than a sound-level survey approach to noise control. Although sound measurement and analysis data should be obtained for use in conjunction with audiometric testing, comprehensive individual monitoring of personnel who move in and out of high-intensity noise fields, which also may vary in intensity, requires a larger expenditure of professional time than most companies are able to provide or are willing to assume.

PRESENT STATUS

Reasons for control of environmental noise are: (1) to reduce hearing losses, (2) to improve communications intelligibility, (3) to increase work efficiency and safety, and (4) to decrease public nuisance. Aside from the humanitarian aspects, the cost of noise

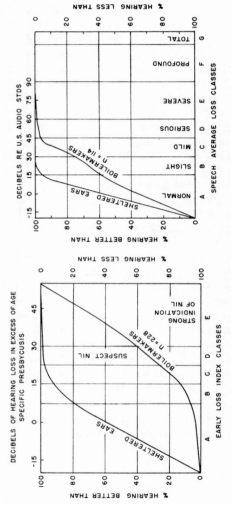

Figure 4. Summation diagrams for comparing ELI and SAL class distributions for boiler-makers and a reference group employed in a large oil refinery.

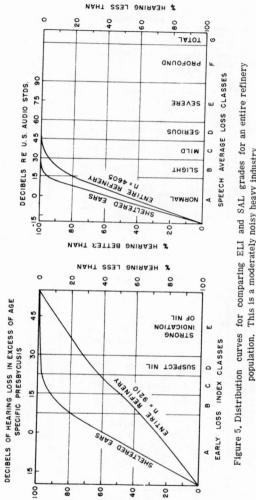

Figure 5. Distribution curves for comparing ELI and SAL grades for an entire refinery population. This is a moderately noisy heavy industry.

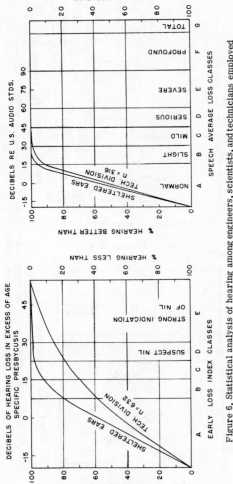

Figure 6. Statistical analysis of hearing among engineers, scientists, and technicians employed in a large oil refinery.

damage to hearing in terms of awards for disability suits and compensation claims is quite tangible. Most notable in the history of compensation in hearing-loss cases was the filing of 232 claims against the Bethlehem Steel Company by workers who claimed more than 5 million dollars in damages. Although initially filed at common law, under the Longshoremen's and Harbor Worker's Compensation Act and the New Jersey Workmen's Compensation Act, settlement was made under jurisdiction of the New Jersey Workmen's Compensation Act after relatively minor litigation. The payments totaled only $ 250,000 excluding lost time, medical fees, and litigation costs. The noise and hearing aspects of compensation laws vary widely from state to state. Recent experience indicates that awards commonly range from $1000 to $3000 per claim. It is unlikely that these rates will decrease in the future.

Industry has borne the brunt of hearing-loss claims even though much of environmental noise exposure and loss may arise from avocational and recreational pursuits. A number of audiometric studies regarding the relation of occupational noise to hearing have been published, of which a few are cited [2-5]. A review of the literature has not revealed similar significant studies relative to nonoccupational environmental noise exposures. Although noise surveys and evaluations of automobiles, subway trains, and aircraft have been made, acoustical effects relative to correlated cumulative hearing threshold shifts have not been defined. Accurate and sensitive audiometric tools and techniques are presently available for such research. We now need devotion of time, money, and personnel to environmental health research in the field.

It has been shown by a comprehensive study of one moderately noisy industrial complex that some 20% of the employees had significant hearing losses in the speech-important frequencies. Indications were that most of this loss was noise-induced. If such a figure is at all representative of hearing status among the nation's workers, more than 14 million industrial employees have hearing impairments to an extent that is compensable in several states. Fortunately, most of this hearing impairment is of a mild to moderate degree and noise-induced hearing loss is self-limiting as shown by Glorig and Davis [6] and Hermann [7]. Among particularly noisy crafts, a speech average loss (SAL) grade of B or less may occur in more than 40% of personnel so employed. High-frequency losses as indicated by early loss index (ELI) grades of D and E may occur in more than 75% of the personnel. Although such hearing decrements are irreversible and cannot be cured by any known medical technique, the noise-induced portions of these losses are preventable. Acceptable degrees of environmental noise control will have to be developed.

REFERENCES

1. Committee on Hearing and Bioacoustics—National Research Council. Handbook of Noise Control. "Scale of degrees of hearing loss," p. 6-2, Cyril N. Harris, (Ed.), McGraw-Hill, New York, 1957.
2. Hermann, E. R., "An audiometric approach to noise control," American Industrial Hygiene Association Journal, Vol. 24, p. 344, July-August 1963.
3. Yaffe, C.D., and Jones, H. H., "Noise and hearing—relationships of industrial noise to hearing acuity in a controlled population," Public Health Service Publication No. 850, U. S. Gov. Printing Office, Washington, D. C., 1961.
4. Riley, E. C., Sterner, J. H., Fassett, D. W., and Sutton, W. L., "Ten years experience with industrial audiometry," American Industrial Hygiene Association Journal, Vol. 22, p. 151, June 1961.
5. Schneider, E. J., Peterson, J. E., Hoyle, H. R., Ode, E. H., and Holder, B. B., "Correlation of industrial noise exposures with audiometric findings," American Industrial Hygiene Association Journal, Vol. 22, p. 245, August 1961.
6. Glorig, Aram, and Davis, Hallowell, "Age, noise and hearing loss," Annals of Otology, Rhinology, and Laryngology, Vol. 70, p. 556, 1961.
7. Hermann, E. R., "An epidemiological study of noise," Presented before the XIV International Congress on Occupational Health, Madrid, Spain, September 1963. Abstracts of Free Communications, Vol. IV, No. 119, E38, Excerpta Medica Foundation.

Odor in Man's Environment

Richard L. Kuehner

Borg-Warner Research Center
Des Plaines, Illinois

We have to admit that of our primitive functions, the sense of smell, like that of taste, has assumed a secondary role. Certainly the anosmics now survive and reproduce their kind with almost none of the limitations of those lacking the sense of sight, hearing, or touch. Without gainsaying the fact that countless lives have been and will be saved by the odor of illuminating and war gases, or that countless dollars have been and will be saved by the odor of overheated motors and burning buildings, a given individual can go days on end without making more than hedonic use of his nose.

In a philosophical sense, man's "being" is a sum total of his past experience and his present stimulus reception. Odor perception is part of this "being." Certainly the sense of smell is used in the enjoyment of, if not in the identification of, and search for food. Certainly, via perfumes, it is still involved in species perpetuation if not as directly as in earlier times. Examine any civilization; the more removed it becomes from day-to-day survival, the more it searches for the complex (the different?) in its sense stimulation. Look at the visual art forms, musical sounds, cooking recipes. Just so is our society pleased by the subtle, the different, the exotic in olfaction. Soap shall smell of spice, or at least "clean"; cars shall smell "new," that is of adhesive and vinyl; materials of construction shall not smell.

In this latter vein one might propose a general rule that unless odor elicits a pleasurable response, there should be no odor. While 30-year-old work at Yale attributes a purely physiological effect to malodors, our major concern with odors beyond their pleasurable use has been to avoid emotional or psychological affront to the sense.

For example, putrescine or cadaverine means death and corruption; valeric acid indicates uncleanliness; skatole, indole, ammonia signify excreta. All of these and many others universally conjure up mental images which are personally offensive, or which imply a less-than-acceptable way of life, living or working environment, and/or inferior socio-economic status.

To avoid this imagery, a surprising amount of money is spent yearly in increasing amounts by our society. In 1963, sixty million dollars were spent by the housewife on scented aerosol bombs to conceal the real or fancied existence of malodors; no proper air recirculating system will return air from bathrooms or kitchens; the air-conditioning industry is paying a market restricting price in pumping outdoor air into occupied spaces to reduce internal odor levels; holes are punched in otherwise entire walls and roofs to vent kitchens and baths; investment in reactors for rendering plants, catalytic and noncatalytic combustors for the organic industries, and chemical and nonchemical absorbers for general purpose deodorization is more than substantial.

As to control of malodors a number of methods are available. It will be as true tomorrow as it is today that source isolation or removal is the most effective method. All else is a crutch. Perfume or chewing gum does not substitute for poor personal hygiene; urinal cakes are no substitute for soap and water.

We are seeing an increasing use of odorless or odor-controlled materials of construction, but where elimination of the source is not possible, as in cooking, smoking, or organic industries, a number of controls are available.

Classically, dilution with tropospheric air is the most common. One method is the collection of odor followed by the stack or exhaust discharge of odor concentrates at a point so removed from sensors that the odor becomes diluted to acceptable levels by convective and diffusive forces before it reaches the sensor. The other is to pump tropospheric air into an odor-containing space to dilute the odor in that space. Both practices are suffering from the same ill. Most air close to man's population concentrations, i.e., his point of need, is becoming so increasingly polluted that, in the one case, he is being prohibited by law from polluting it further or, in the second case, his ventilation air brings in contaminants more offensive than the odors he thereby wants to dilute.

In anticipation of this loss of ventilation air, alternative control methods have been and are being developed. One of these is masking. Masking can take several forms all of which involve the addition of something to odorous air or odorous material which make the malodor less observable without altering the actual concentration of the malodor. This can be a pleasant perfume which either overpowers the malodor or blends with it to yield a more acceptable complex. Or it may be a specifically irritant or anesthetic gas which either temporarily or permanently depresses the sense of smell.

A third method is to collect the odors in ad- or absorbent liquids or solids. In this instance the odorous air must be brought in intimate contact with the sorbent; the sorbent must be discarded or regenerated once it has ad- or absorbed its full complement of odor.

A fourth and final method is the destruction of odor by chemical reaction, usually oxidation. Chemicals which are powerful enough to react with odors will also react with biological systems. Therefore, it is normal to support such chemicals on a solid or liquid medium which will accumulate the odors and bring them and the reagent chemical in intimate contact. In this method the reagent chemical, once spent, is replaced with or without replacement of the supporting medium.

This has been a thumbnail sketch of odors and odor control. It is a science which has contributed to man's survival, and one which is increasingly contributing to a complex society's health, pleasure, and sense of well-being.

The Sensory Environmental Spectrum

Max V. Mathews

Bell Telephone Laboratories
Murray Hill, New Jersey

In dealing with our sensory environment, our first concern is with the sounds of our times. The sounds have changed in the last fifty years from those of a rural civilization to those of an urban civilization, and in many ways we don't realize how much they have changed because we didn't record the sounds in previous periods and we can't really listen to them. Composers with sensitive ears in some sense have recorded the sounds. It has been said that the music of Mozart, for example, reflects the buggy wheels going over the cobblestones of the streets in the city in which he lived. In the same sense, perhaps, the music of modern composers, such as John Cage, reflects the sounds of jet planes, automobiles, and machines, and I think you have all heard this music, whether or not you enjoy it.

A previous speaker mentioned the hearing loss at high frequencies from which we now suffer. We might well ask how much of this hearing loss is due to our acoustic environment and how much is due to the natural process of aging, and this is a question we must answer shortly.

Not all the effects of civilization are objectionable as far as sounds via the telephone which didn't exist sixty years ago and was impractical twenty years ago. Now it is really possible to work for reproducing sound with a very high degree of fidelity, using loud-speakers and electronic amplifiers. We have very convenient and compact ways of recording sound, principally the high-fidelity phonograph record; consequently, we now have easily available in our own homes an enormous selection of musical sounds and, for that matter, almost any other sound you might wish to hear. It is cheap and has given to the entire population a range of music which was only enjoyed by a very few people some years ago. We also have magnetic tape recording and playback techniques. This has not improved the fidelity that one achieves in sound, but it has improved the convenience of recording and patching together sounds, and it has also meant that anyone can record and save sounds.

Furthermore, nowadays we have a range of communication with sounds via the telephone which didn't exist sixty years ago and was impractical twenty years ago. Now it is really possible to work anywhere and have a telephone there and communicate with the people you wish to communicate with in order to carry out your work. Your home can really be your office, if you wish it to be.

Now let's consider light and the control we have of light. I would mention something very obvious here: electric illumination has completely changed our lives. In the illumination of houses, for example, we can now work at night, or read at night, as well as in the daytime. With the illumination of automobiles we can drive effectively at night. In the last few years we have been getting into the illumination of sports arenas so that we can take our recreation at night.

I would also mention television as a control of light and say that here we now have a medium which is able to preempt or saturate, if not our entire sensory input channels, at least a sufficient portion of what the brain wishes to concern itself with, so that we no longer need to get bored. In this sense, I think television is a different creature than radio was. I think you can entertain people endlessly with this medium. Whether this is a good thing or a bad thing, I'm not sure, but it is, at least, the current equivalent of the old Roman circus.

Individual communication of pictures and voice, such as the picture phone, is now technically possible but still very expensive and not widely used.

I've mentioned sound and light. I would like to end by mentioning one thing which I'm not sure is sensory, but which is very important, and this is the control of information. The printing press is still the main communicator via books, magazines, and newspapers, but the cathode-ray tube is now beginning to be a very close competitor. The cathode-ray tube is in the television set, and soon the cathode-ray tube connected with the computer will, I think, indeed replace the printing press. This will allow individual interrogation of computers via some communication channel in order to get answers to individual questions.

The Relationship of Man to Microbe

James G. Shaffer, Sc. D.

Lutheran General Hospital
Park Ridge, Illinois

In the economy of nature, it is impossible for any species of plant or animal to exist without exerting effects on other plants or animals. Microbes (bacteria, protozoa, and fungi), being perhaps the earliest forms of living things on earth, have exerted their influence in a multitude of ways, playing decisive roles in the survival of species, in the control of populations, and in the overall determination of the contour of life on this planet.

Man's relationship with microbes began with Adam and has evolved through the ages into a complex interplay which, until relatively recent times, remained a great mystery. The broad range of man's association with microbes covers all the elements of his existence on earth. In terms of health, it runs the gamut of the delicate balance of the host—parasite relationship to the violence of the acute, fulminating disease process which may result when the balance is overturned. Various microorganisms exert effects on man's food supply, either directly or indirectly. Microbial diseases in cattle, swine, sheep, poultry, grain, vegetables, and fruit account for enormous economic loss and disruption of food supply. On the other hand, many microorganisms supply essential elements which either directly or indirectly affect man's well-being. It is interesting to note that man has been better able to cope with his larger associates, i.e., ranging from the dinosaur to the mouse, than he has with his smaller associates, i.e., ranging from the insect to the microbe.

There has evidently been an awareness of a relationship between man and microbe from the earliest times of recorded history. The practice of inoculation for smallpox developed by the Chinese, certain rules laid down by Moses, the writings of Hippocrates, and various references to be found in the Bible bear testimony to this. Throughout the Dark Ages, there is to be found a thread of belief in the infectiousness of certain diseases such as smallpox and plague, but it was not until the advent of Pasteur and Koch and their contemporaries that the relationship of man and microbes was brought into focus. Although Pasteur is best known for the elucidation of the "germ theory" of disease, it is important to note that his work with the wine industry led to the basic understanding of fermentative processes, saved an industry, and led to the development of new facets that are now indispensable. His work with silkworm disease and with certain of the animal diseases also stands as a great contribution to the development of our knowledge of man's relation to microbes.

Following Pasteur's original work, the greatest impetus for the study of microbes came from the desire to conquer disease. During the latter part of the 19th century and the early part of the present century, the great scourges of mankind were studied intensely. Epidemiologic investigation revealed most, if not all, of the basic information needed to control the acute fulminating, contagious diseases that had traveled in epidemics over the world for thousands of years. Application of the basic principles that were evolved in these studies led to great strides in the control of such diseases as smallpox, typhoid fever, dysentery, plague, diphtheria, whooping cough, malaria, and others. When, in the late 1930's, the sulfa drugs were introduced and were then followed by the antibiotics, the conquest of most of the acute, contagious microbial diseases was literally complete.

With the battle against the great killers won in large areas of the world, it is now possible to afford the luxury of studying in some depth the ecology of the relationship of man and microbe. The history of man on earth has been mightily affected by disease. His gregarious nature and his tendency to aggregate in villages and form armies for conquest have resulted in the concentration of people under what have, until recent times, been, more often than not, unsanitary conditions. From the vantage point of our 20th century knowledge, we can see that the close association of people under these circumstances was fertile ground for the genesis of epidemics in much the same way that a dense forest provides the setting for a conflagration set off by a single spark.   As travel increased, the spread of epidemics from village to village and from country to country led to the propagation of real disaster.   Smallpox, for instance, swept across Europe in waves, periodically decimating populations, each sweep ending when the susceptible population was reduced so that insufficient contacts occurred. Plague, cholera, influenza, diphtheria, to name a few others, also took their toll.   Political or military success or failure not infrequently was determined by the impact of disease; often dysentery or plague.   Thus, military heroes were on occasion, perhaps, generated as a result of disease in the enemy ranks, not by the development of an unbeatable strategy.   In the settlement of the North American continent, the success or failure of a new colony was often determined by the ability of the colonists to avoid the ravages of disease. Following successful establishment of the original colonies, the advance of white settlements was made to considerable degree against resistance from the native Indians.   This resistance was reduced by the ravages of the white man's diseases, especially smallpox and probably measles. Whole villages were wiped out, others decimated, and it seems entirely likely that more Indians were destroyed by the white man's diseases than by his bullets.   Some areas of the world are, as yet, not settled by humans because of the presence of certain uncontrolled diseases.

There are still large areas in the less developed and densely populated parts of the world where microbial disease takes a heavy toll of life, reduces efficiency, and accounts for enormous economic loss. The means of accomplishing control of such diseases as malaria, tuberculosis, smallpox, and most others that exist in these regions are known and accomplishment only awaits the application of proper measures.

It would be wrong to leave the impression that the reduction in disease signifies elimination of the microorganisms that cause these conditions. Many of the organisms are still present, constantly being transmitted from person to person but prevented from assuming epidemic status to a degree by a combination of environmental control, vaccination, and appropriate medical treatment.   In some cases, however, the organism has been eliminated. For example, smallpox has been nonexistent in the United States for a number of years, and it seems possible that one might hope for the elimination of the polio- myelitis virus with extensive use of the oral vaccine.

Let us now turn to the intimate relationship that exists between man and what has been termed the normal microbial flora. There exists upon all surfaces of the skin, in the mouth and the throat, and in the lower portions of the digestive tract a complex and variable complement of microorganisms.   It is the rule rather than the exception to find in this flora organisms that are capable of causing disease.   Study of this situation leads one to postulate that the normal relationship between man and these microbes is a harmless host—parasite existence.   We can term this an infection, which, under circumstances that are as yet unknown, may be converted into disease.   Thus, the great unknowns in man's relation to microbial disease are the complex determinants that allow disease to develop from infection.   What is implied here is that the infected individual is in great measure more important than the microbe, for it is the host that provides the milieu for the occurrence of disease. Ironically, the success that has attended the application of public health and medical measures to keep people alive results in the development of an increasing number whose susceptibility to microbial disease is enhanced. Indeed, we now find diseases being caused by organisms not considered as pathogens in earlier times. This, however, does not portend the outbreak of new epidemics, unless unforeseen mutations occur. However, it is possible that we are developing, to a degree, an antibiotic-dependent populace.

At the present time, there still exist circumstances under which epidemics can occur. Army posts where large numbers of new recruits are brought together for training are potential sites for epidemics, as witnessed by the recent outbreak of meningitis at an army camp. Our large urban areas are potential sources. Slum clearance, city planning, and the design of our magnificent multistory office buildings and apartments must be planned to avoid undue concentrations of people and microbes. It should be pointed out here that, until relatively recent times, cities were unable to maintain their population without an influx of the more aggressive people from the less disease-ridden rural areas.

The childhood diseases such as mumps, measles, and chickenpox still exist in epidemic form. These are caused by viruses, as are influenza, infectious hepatitis, the common cold, and a number of other milder conditions. So far, no real control of these conditions has been established, although the recently developed measles vaccine is promising, and there is hope that a hepatitis vaccine is on the way. There is now a strong suspicion that at least some forms of cancer are caused by viruses, and the possibilities that this suggests are indeed intriguing. This suggests a type of intimate relation between the disease-producing agent and the tissue cell not yet observed with other agents.

It would not be proper to end this discussion without a look at the activities of microbes that are beneficial to man. The suggestion has been made that the normal flora of the intenstine, especially that which exists in the large bowel, contributes certain vitamins. Certainly many of the bacteria in this region produce vitamins of the B complex, but the evidence for their absorption into the body is not conclusive. The growth of animals in a totally germ-free environment would seem to indicate that the contribution of the microbial flora, whatever it might be, is not essential.

In the economy of nature, the nitrogen-fixing bacteria play an important role in the maintenance of soil fertility, and the putrefactive organisms are an essential part of nature's scavenger system, destroying organic debris and making it available for utilization by other living things. Use is made of these microorganisms in sewage disposal systems throughout the civilized world. In fact, were it not for the fact that microbes have the ability to break down any and all organic wastes, man might find himself wading around in a muck of dead organic matter.

The fermentative properties of microbes which have been alluded to before are utilized by man in numerous ways. The production of alcoholic beverages depends on this exclusively, although when one considers the problems that alcoholism presents to society, it is obvious that this is not an unmixed blessing. On the other hand, one suspects that without alcohol the afflicted person would find some other means of exhibiting his symptoms. A number of the organic solvents used by industry are produced by microbes under appropriately controlled conditions.

It is always intriguing and perhaps dangerous to attempt to look into the future. Already certain microorganisms are being studied as sources of food for space travel and for the population, if and when it reaches the point where sufficient food cannot be produced from the soil. The dual role of regulating the atmosphere in space vehicles and providing a supply of food is an interesting thing on which to speculate. In another area, experiments are in progress, designed to see whether or not microbes can be used to control insect populations. These are exciting ideas, but full of considerable hazards, as is well illustrated by the experiences in Australia with the introduction of the European hare and subsequent attempts to control it with the virus of infectious myxomatosis. Man's attempts to control the ecology of nature often lead to the creation of new and more complex problems.

In fact, our ability to prevent disease and prolong life is in no small way responsible for the population explosion about which so much is being written. One wonders what will happen when, in a nation like India, with its high birth rate, the application of modern health practices reduces the infant mortality rate and extends life expectancy to that which exists in the more advanced nations. Thus, the solution of one ecologic problem often results in the production of new ones, more challenging than the one that was solved.

That we are capable of controlling the environment successfully and solving the problems developed in a modern world, at least as they apply to epidemic disease, is indicated by the success which attends the operation of hospitals. Here one has all the seeds of disaster. There is concentrated under one roof a group of highly susceptible patients into which, periodically, there are introduced individuals with disease caused by the most dangerous disease-producing microorganisms. In many ways, this environment represents all the hazards, both old and new, that would conspire to initiate an epidemic. Yet, with the application of relatively simple techniques, all of which can be easily documented, the hospital provides a safe environment for the care and management of human ills and does not constitute a source of contagion for those in the hospital or in the

surrounding community.   Much can be learned from studying the
hospital environment that will be of value to those concerned with
community problems.

Microbes are capable of mutation and change. What new properties
will they assume? Man is an even more complex and changeable entity.
What new relationships between man and microbe, either beneficial
or harmful, will develop in the future? It will be interesting to see.

Discussion

QUESTION (Professor Cember): What do you foresee in terms of environmental control of viruses?

SHAFFER: Many of the measures that apply to the control of bacteria also apply to the control of viruses, although these agents are to a degree more resistant to some of the measures than are the bacterial agents. Fortunately, they often produce much milder conditions than the bacteria, and many of them merely lay the host open to bacterial invasion. Perhaps vaccination in many of these situations is the best means of control. We're considering right now trying to air-sample our hospital to determine viral content of the air. The danger of transmission of respiratory viruses within a hospital environment has not yet been evaluated as thoroughly as one would like. Much more information is needed before decisions can be made about any special control measures.

QUESTION: I would like to ask Dr. Kuehner if he would comment on alumina—permanganate odor control, and, very briefly, on the value of the oxidation process for destroying odor. This is something fairly new.

KUEHNER: Historically, since World War I, there has been only one real synthetic odor control and that is activated carbon, which was developed for gas masks. This is a very effective product, albeit expensive. It has certain disadvantages. It stores odors. Anything that is stored can be unstored. Carbon, under certain conditions of temperature and humidity, can desorb part of the odor it has absorbed.

The product to which you refer combines carbon's absorbitive properties, although alumina is the basic material. But rather than permit the odors to store up, they are attacked by an oxidant, the preferred being potassium permanganate. Potassium permanganate is preferred for this reason. There are a number of oxidants that can take odors from odorousness down to odorlessness as, for example, chlorine.   In so doing, chlorine produces intermediate products; intermediate stages which are malodorous.   Tobacco smoke reacts with chlorine and you get a frightful smell before it goes to odorlessness.   Permanganate has the ability to go in gradual steps down to odorlessness, none of which are malodorous.

QUESTION: Dr. Mathews, have there been any studies of what an optimum noise level and spectrum is for the various kinds of activities such as doing arithmetic, reading?

MATHEWS: Yes, there have been a number of studies, particularly in England, by a psychologist by the name of Broadbent, and the results are rather peculiar.   He studied how much degradation in doing a particular task, either driving a car or doing mental arithmetic, was produced by various degrees of noise in the environment, and it turns out that almost no degradation was introduced... let's say by noise alone, until the subject was put under some additional stress... some third factor, which could be extreme fatigue or something else, so that in some sense his mental capacity is really saturated.   We seem to have a reserve capacity which, although it is aware of the noise and is disturbed by it, is able nevertheless to fend it off so that we can cope with the job at hand.   I can't quote you the exact results here; you should probably look up some of his work if you are interested in this question.

QUESTION: Dr. Shaffer, is there any relationship between the ability of an environment to transfer bacteria or viruses and its odor level?

SHAFFER: We're almost going back to malaria here.   Malaria really means "bad air."   The presence of odors might imply that there is a possibility for bacterial contamination of such air under certain circumstances.   I know of no studies that have been done in this sort of area.   However, we're more concerned about dust and other particles large enough to have bacteria attached to them than we are about odor.   Odor, by and large, doesn't have a direct relationship with microbes except that they may produce odors when they putrefy dead organic matter.   It is possible to have large numbers of bacteria in the air, without there being any odor present.

QUESTION: I was referring to the mechanisms rather than relationships—mechanisms for carrying odor.   Does this mechanism carry bacteria and viruses?

SHAFFER: Yes, it is possible.   I know of no data, but I imagine that the same mechanisms would function. Many of the measures that one would take to control the spread of gases or odors in the atmosphere might also have the effect of controlling the transmission of microbes. Thus, our activities in both areas are probably related.

KUEHNER: On your question, keep in mind that in the case of odors you deal with a gas and you have diffusive phenomena taking place; with bacteria these are particulate matter, in which you have only convective currents supplying motion. Secondly, with bacteria you have one advantage in control, in that you can kill them and you don't have to remove them. With odors you've got to get them out.

SHAFFER: Studies in closed spaces such as simulated operating rooms show that when you drop a bacterial aerosol into a spot, it diffuses in much the same way that odors would, or as a drop of alcohol in a glass of water. There is almost an immediate dispersion of the bacteria throughout the enclosed space. Perhaps it does follow somewhat the same laws as those dealing with the dispersal of odors.

It is the usual thing to think of killing the bacteria. All the evidence that we have as we study this in the hospital, and in other areas, is that removal is much more effective than killing. There are reasons for this. One of the reasons is that microbes are biological and they have the ability to adapt and change and no matter what magic bullet you aim, sooner or later there may appear one that is resistant to a particular substance. Studies done by others have shown that mopping a hospital floor with warm water reduces the bacterial count on the floor to the same degree as mopping it with a disinfectant. In our hospital, we lay considerable stress on removal of bacteria and reduction of bacterial content and have, to a degree at least, de-emphasized concentration on killing with a disinfectant.

KUEHNER: I pass, but without prejudice. I'll argue this later.

QUESTION: Mr. Viessman, do you have any information on human work efficiency under very extreme conditions?

VIESSMAN: Cold or hot?

QUESTION: Cold.

VIESSMAN: Cold. No, undoubtedly as the person becomes un-comfortable his work efficiency slows down. You've got to keep him within a reasonable degree of comfort, otherwise his strength is dissipated in trying to be comfortable.

QUESTION: Heat?

VIESSMAN: The humidity as well as the temperature has a decided effect on high-temperature comfort. This is one approach to heat-stress problems. There are four or five other methods of discomfort or stress analyses or standards, but none of them is absolute. They all have their limitations. Because of time restrictions for presentation, I have limited my discussion to the effective temperature concept.

COMMENT FROM THE FLOOR: We are talking about extremes causing stresses and I wonder if we have any information on whether the optimum conditions might leave distress. I notice in my own work through 25 to 30 years of residential air-conditioning that people are almost automatically, summer and winter, adapted to about 75°F temperature. No wonder the old 70°F in the living room isn't at the summit on the ASHRAE comfort chart, which is a minimum stress point.

VIESSMAN: That's right. The chart indicates the most favorable areas and the temperatures which humans desire in winter and summer. Most people are comfortable in winter with conditions shown by the panel observation peak and that is around 67 to 68°F ET, depending on humidity. In summer, the comfort peak is around 71°F ET. This is just a brief treatment. The comfort chart and a more detailed discussion is contained in the 1965 ASHRAE Guide and Data Book, which shows the newly revised comfort chart. A person is most effective in his ability to perform his tasks when he is comfortable.

KUEHNER: Commenting on the question of stress and temperature, the air-conditioning industry goes by this 75°F. They also go with a 50% relative humidity. We scientists have no objection to the 75°F, but speaking from the odor standpoint, you are wrong about the 50% relative humidity. As you increase the humidity, the sensitivity of the sense of smell is depressed. However, at the same time you increase the rate of generation of odor from linoleum, draperies, etc. For the best balance of depressing this generation of intrinsically odorous substances and decreasing the sensitivity of the sense of smell, 40% is really the best humidity.

COMMENT FROM THE FLOOR: What is really important in a totally controlled environment is that if you can control the temperature, the relative humidity doesn't seem to play an important part. You can live in an area of 25 to 65% relative humidity and not recognize the shift; the humidity range is not important. It's only when you get up to 78 to 79°F dry-bulb temperature that you have this tremendous relative humidity effect.

VIESSMAN: Previous standards of comfort in wintertime have been effective temperatures between 65 and 70°F, and in summertime from 68 to 73°F. Now the new ASHRAE chart has a slight modification to that. The new chart shows that comfort conditions can be experienced at 75°F with humidity ratios of 20 to 70%. This applies equally to summer or winter. At higher humidities, lower temperatures are required for comfort.

COMMENT FROM MEMBER OF PANEL: On this question of extreme conditions, I believe the Army Quartermaster Corps' Research Laboratory has, as far as I know, done more on that than anyone else.

COMMENT (GOTAAS): Once somebody gets into a room at 75°F you can change the relative humidity and he will not know it as long as you keep the room at 75°F and keep him at the same creative activities. We can use a simple thermostat with one setting, leave it at 75°F, and our employees just don't care enough. One thing we now specify is thermostats without thermometers. If it is actually 75°F, they're comfortable, and if they don't touch that thermostat, regardless of humidity, and leave it at 75°F, away we go. We don't always have perfect humidity control, sometimes only one, and it goes on and off. It may go from 25 to 65°F in the summertime. In the wintertime, you can't maintain 50% relative humidity in many buildings because of condensation on the windows. This seems to be the practice; many engineers are doing it this way; they insist that there be no thermometers on the thermostats.

COMMENT FROM THE FLOOR: During the winter months we have a peak in respiratory infections. I would like to know whether it is the feeling of anyone here that giving us this dry, overheated, hot air in the wintertime is in any way responsible for this deplorable condition. If not, what can be done to bring down these high peaks of respiratory infections with control of the atmospheric environment?

SHAFFER: This is certainly one of the possiblities, and it gets back to what was said about the human host being important in determining whether or not disease develops from infection. Certainly drying up mucous membranes must have some effect, and it is likely that this effect has to do with the ability of a microorganism to survive on this surface. You probably know that the fluids of the mucous membrane have some antibacterial effect. If you remove these, you remove one of the defense mechanisms. I would suspect that there are other factors, such as aggregation of people in enclosed areas with poor ventilation and fluctuations in their body temperatures. This may affect the ability to support the growth of microorganisms, i.e., chilling or overheating. One might suspect that there is a relationship between the low humidity that one finds in many places in the wintertime. What that relation is and what the mechanism is would be hard to define, although as you know, under normal circumstances the nasal passages as they go down into the lungs are very efficient at removing organisms inspired in the air. One wonders if this drying out doesn't have an influence on this.

There's another point that should be mentioned here in connection with humidity. It is true that microbes in the air survive better at certain humidity levels than they do at others. Some survive better at low levels, others do better at high levels. One suspects that there is a whole spectrum of relationship between humidity level and survival of bacteria. Perhaps the low humidity in winter provides better survival for certain agents associated with respiratory diseases.

QUESTION: Dr. Shaffer, you mentioned that once in a while you've missed with the magic bullet and I assume you referred to mutations. What is your opinion that the rate of mutants will outrun your supply of bullets eventually?

SHAFFER: That question really can't be answered at this time. Of course you have to remember that microbes have a generation time that may be as short as 15 minutes so that the potential for mutation in the microbe is much greater than it is in the human or the higher animals. This is certainly a possibility. The other thing is the fact that there often exists in the environment organisms already resistant to these magic bullets. These may become established when other organisms are removed. We found one of the reservoir tanks in a humidifier heavily contaminated with bacteria on one occasion. Our maintenance man promptly tripled the amount of disinfectant that he was using in this tank. The next day we went back and cultured the water in the tank, and there were more bacteria that day than there were before. Evidently, the bacteria were not the least affected and may have been encouraged by the added disinfectant. This is not an isolated observation. Other people have seen the same thing. Sometimes it is a good idea in environmental control to change disinfectants occasionally. Whether by mutation or whether by the introduction of organisms that are already resistant, the disinfectant in use loses its effectiveness.

The Environmental Spectrum—Tomorrow

Moderator: Richard L. Kuehner, *Manager, Environmental Sciences, Borg-Warner Research Center, Des Plaines, Illinois*

Ionizing   Herman Cember, *Professor of Environmental Health*
Radiation: *Engineering, Northwestern University, Evanston, Illinois*

Medical: Captain G. F. Bond, *Medical Corps, United States Navy, Washington, D. C.*

Pollution   J. E. Quon, *Associate Professor of Environmental Health*
Control: *Engineering, Northwestern University, Evanston, Illinois*

Toxics: Theron G. Randolph, M.D., *Staff Member, The Swedish Covenant Hospital, Chicago, Illinois, and Lutheran General Hospital, Park Ridge, Illinois*

Urban Area: Frederick T. Aschman, *Executive Vice-President, Barton-Aschman Associates, Chicago, Illinois*

Ionizing Radiation

Herman Cember

Environmental Health
Northwestern University
Evanston, Illinois

This session is devoted to speculation about environmental health problems of the future and specifically, in my presentation, to the environmental health problems arising from man's use of ionizing radiation and nuclear energy. We speculate about the future by projecting our past experiences in the light of our present knowledge. The further into the future that we project, the greater is the likelihood of being wrong. In the case of radiation, there have been many divergent opinions and much speculation about the possible problems in public health that may arise as a consequence of man's full exploitation of ionizing radiation and nuclear energy. I would like to air my views about this point.

Naturally occurring environmental radiation has been with us, we believe, at least since the days of Adam and Eve. Despite the fact that people had been suffering this continuous, low-level radiation insult since the beginning of time, no direct etiologic relationship between this radiation insult and any known disease was immediately observed. (A direct relationship has been established between the lung cancer suffered by the miners in the Joachimsthal and Schneeberg mines and the atmospheric radioactivity in those mines. In that case, however, the radioactivity levels were very high relative to our present safety standards.) Cases of known injury from radiation occurred after the discovery of X rays by Röntgen in 1895 and radioactivity by Becquerel in 1896. The first such case, within one month of Röntgen's discovery, was the production of erythema of the hands of a Chicago physicist called Grubbe. Becquerel, himself, suffered radiation injury that required medical treatment. Carrying a glass vial of a radioactive salt in his vest pocket for several days resulted in an ulcer on his chest. He soon found that wrapping the vial in a sheet of lead sufficiently attenuated the radiation to permit carrying it without producing visible signs of radiation injury. The early experiences with radiation sources, as well as the long history of exposure of the entire world's population to the relatively low-level naturally occurring radiation, showed that radiation was not dissimilar from the other noxious agents with which man deals. The crux of the problem in radiation safety was the quantitative aspects of the dose—response relationship. The question that we must ask is not "Is radiation damaging?" but rather "How much radiation damage is acceptable?" Although much still remains to be learned about the biological effects of radiation, enough has already been learned about the answer to the second question to permit the safe use of radiation sources. On the basis of this knowledge, neither industrial nor medical uses of radiation either now, or in the fore-seeable future, are expected to pose any serious problems in environ-mental health. The main source of potential environmental health problems from radiation will be, I believe, the expanded use of nuclear energy.

Most of the experts on energy resources agree that mankind will be forced to depend ever more strongly on nuclear energy as our present reserves of fossil fuel and other natural resources become more depleted and more uneconomical to exploit, and as our population continues to explode. In the production of nuclear energy, enormous quantities of highly toxic radioactive wastes, called fission products, will be generated. The nature of radioactive waste is qualitatively different from that of other toxic industrial wastes. Nonradioactive wastes can be disposed of after undergoing suitable treatment to reduce their toxicity. In the case of radioactive waste, nothing can be done by technological means to reduce the radioactivity, and, hence, its toxicity. Only time, during which the fission products decay naturally, can decrease the radioactivity. Small amounts of radioactive waste can be discharged into the environment, there to be dispersed and diluted to very low levels that are considered acceptable. In the case of the high-level radioactive waste that is produced during the generation of nuclear power, the total physical environment, including the atmosphere and the seas, just isn't big enough to dilute all the fission product wastes to levels that are considered safe. The only practical alternative is to store these high-level wastes, for hundreds of years, until the radioactivity decays. Such storage, of course, must be under extremely rigorous conditions of security. At the present time, storage presents little difficulty. However, as nuclear power generation continues to increase, the management of the inevitable large quantities of high-level wastes is expected to tax our technological ingenuity.

In setting acceptable limits, a large number of biological effects were considered; those of greatest concern in the case of very low-level exposure to the total population of the world are life-shortening, carcinogenesis, and genetic effects. The life shortening, I believe, can be easily discounted. Even if we take the worst extrapolations from mice to men (there are not, at this present time, any data that clearly show a life-shortening effect on man, so we must extrapolate from animal data), the most pessimistic guesses predict life-shortening effects of the order of days per lifetime. This, it seems to me, is quite insignificant when one considers the fact that a heavy cigarette smoker "enjoys" a life expectancy decrease of about seven years over a nonsmoker.

The second effect of concern is carcinogenesis or, more specifically, leukemogenesis. The relationship between exposure to radiation and increased risk of leukemia is rather well established. This relationship, which is based mainly on high exposure doses, can be extrapolated to the low levels of exposure that may be experienced under our present radiation safety criteria. If the entire population of the world were to receive the maximum acceptable dose, and, if we were to make the most pessimistic extrapolations, we would predict one to two cases of radiation-induced leukemia per 17 million population. This is to be compared with a "spontaneous" leukemia frequency on the order of 60 per million population per year. It would be extremely difficult, if at all possible, to actually identify radiation-induced leukemia under these conditions. These figures on leukemia emphasize, I believe, one of the real differences between environmental health problems of the past and the environmental health problems of the future. In the past, we were called on to control environmental hazards in which there was a high probability of producing a deleterious effect to a relatively small group of population at risk. In the new era of public health that we have already entered, we must continue to control these high probability hazards, but we also are faced with the new problem of controlling the very low-risk hazard in which the size of the exposed population is very large. Low risk, as used here, means that the person exposed stands a very low likelihood of being adversely affected by his exposure. Lung cancer from cigarettes might be cited as a low-risk hazard. The difference between the "old" and "new" public health is not merely a difference in degree; it is also a difference in kind. In the high-risk type of exposure, such as typhoid fever or malaria, the effect of the harmful agent could clearly be identified. In the low-risk type of problems, such as air pollution, radiation-induced leukemia, etc., it is either extremely difficult or impossible to identify an individual who unequivocally can be said to be suffering as a result of exposure to the harmful agent.

Included in this general area of low-risk hazards is the third possible result from exposure to ionizing radiation that must be considered: genetic effects. Possible genetic effects, we believe, may have a significant effect on the population, and hence are used as the radiation safety criterion for large population groups that may be exposed to low levels of radiation. It should be emphasized that the exposure levels about which we are talking are so low as to preclude any reasonable likelihood of finding somatic radiation damage in anyone. Geneticists estimate that if everyone of reproductive age experienced a radiation dose somewhere between 30 and 80 roentgens, the present spontaneous rate of genetic defects would double. However, this doubling of the mutation rate would take place over a very long period of time. The half-time for the increased mutation rate, that is, the time during which one half the predicted increase would occur, is on the order of six to seven generations, or about 200 years. Another 200 years, or a total of about 400 years, would be required for the mutation rate to increase halfway again to the predicted doubling of the present rate; this would bring us up to a mutation rate of 75% greater than the present rate over a period of the next 400 years. It should be pointed out that these predictions are based on all members of the population receiving a "doubling dose" during their reproductive lifetimes, and that this rate of exposure be continued for many centuries. At the present time, thanks to our engineering controls over all the processes involving radioactivity, the population at large is receiving only a very small fraction of the maximum acceptable exposure; it is expected that this state of affairs will continue into the immediate future. Thus, our present level of control, if projected into the immediate future, is expected to maintain the genetic mutation rate at a level that is not significantly different from the spontaneous rate of the prenuclear era. From this discussion, it is clear that we have a period of grace of many years before the magnitude of the environmental health hazards associated with the disposal of high-level radioactive wastes exceeds our technological capability to deal with them.

The first principle in dealing with a hazardous substance is to eliminate the hazardous substance whenever possible. Such a step will, I believe, not only be possible, but will be taken before the generation of nuclear energy poses a serious threat to future generations through environmental contamination. The present method of harnessing the energy of the atom is based on nuclear fission, which, as was pointed out above, produces highly radioactive wastes. Another way in which energy may be extracted from the atom is by nuclear fusion, in which two very light nuclei are driven together to form a heavier nucleus whose mass is less than the sum of the two reactants. This loss of mass appears as energy. A unique characteristic of the fusion process is that no radioactive fission products are formed. Energy is now extracted from matter by the fusion reaction only in the hydrogen bomb, which is an uncontrolled reaction. We have not yet learned to control the fusion reaction as we have the fission reaction. The problem of controlled fusion is a major research effort of about 1700 scientists throughout the world. At this time, we still do not know how to attain control. However, 15 years of research effort has not uncovered any fundamental physical reason to show that controlled nuclear fusion is not feasible. When this scientific breakthrough is achieved, an essentially infinite source of nuclear power will be tapped which will not contaminate our environment with radioactivity.

The Medical Environmental Spectrum

Capt. G. F. Bond

Medical Corps
United States Navy
Washington, D.C

It seems to me that we have been talking at considerable length of the efforts of the engineers to give us in our closed environments the equivalent of clean, fresh sea air. I submit that this is possibly not the best of all gaseous media for man to breathe, and certainly under the conditions in which I foresee working in the future, this is not the case. We have been looking for the past several years at the feasibility of putting man as a free agent on the ocean bottom to a depth of at least 600 ft, there to live under ambient pressures and to do useful work on the continental shelves of the world. To that end, it is necessary to go to a synthetic gas environment and the "ball park" changes very, very radically. In the three minutes that are left to me, I'll tell you how it changes.

Last fall, after seven years of work, we placed four men on top of a submerged volcano approximately 200 ft deep. They lived in a habitat about 40 ft long and 10 ft in diameter, from which they were free to come and go on the ocean floor. The aquanauts stayed in this environment for a period of 12 days and were returned safely to earth—to the upper earth—with no subsequent ill effects. We now intend to expand this program considerably.

The environmental conditions under which these people have to live are about as follows: At 200 ft, we must reduce our percentage of oxygen rather radically. We must replace nitrogen with helium and do some other tricky things with this atmosphere. My men were breathing a mixture of 16% nitrogen, 4% oxygen, and 80% helium. This provided the proper partial pressure of oxygen for the aquanauts. They got along very nicely. They lost body heat at a great rate, as you might have guessed, because of the high thermal conductivity of helium. We usually have to increase our environmental temperatures up to about 91°F at this depth and will have to go higher as we go deeper.   There are many other problems in addition to that of heat loss, one of which is the difficult of communication in the helium atmosphere. This, I think, we have finally solved.

Finally, we have some rather interesting psychological problems of closing people up, and exposing them to an extremely hostile environment without much control from us. We are working on these, and I can only promise that within a matter of a few years we will be capable of placing a large number of men to work freely on the ocean bottom at a depth of 600 ft.

Air Pollution Control

J. E. Quon

Environmental Health Engineering
Northwestern University
Evanston, Illinois

The major task of the panel on "The Environmental Spectrum—Tomorrow" is to delineate probable changes in the characteristics of the air pollution control problem in the future on the basis of present knowledge. In this regard, it is well to keep in mind the speculative nature of any projection, and that the degree of uncertainty increases exponentially with the time span. The definition of air pollution given by the Engineers Joint Council is:

> Air pollution means the presence in the outdoor atmosphere of one or more contaminants, such as dust, fumes, gas, mist, odor, smoke, or vapor, in quantities, of characteristics, and of duration such as to be injurious to human, plant, or animal life, or to property, or which unreasonably interfere with the comfortable enjoyment of life and property.

The relationship between the use of fossil fuels and air pollution was recognized as early as the 13th century. Use of fossil fuels, in ever greater quantities to satisfy the energy demands associated with industrialization, higher standards of living, and urbanization, results in the discharge of pollutants (both primary and secondary) at a rate sufficient for the concentrations to reach levels considered adverse to the public welfare. In spite of this early recognition, major efforts devoted to the understanding of the basic characteristics of the problem and control of air pollution were not forthcoming until after World War II. Since then, we have learned a great deal about how to cope with certain aspects of the problem, e.g., dustfall. Many other aspects of the problem remain unresolved. Two difficult areas singled out by the Clean Air Act of 1963 are automobile exhaust control and the economical removal of sulfur from fuels.

127

Air pollution as we know it today is intimately associated with energy transformation involving fossil fuels. As our technology of other energy transformations develops, the energy demands associated with industrialization, with a higher standard of living, and with urbanization may be satisfied in part by a variety of energy sources not dependent on fossil fuels, such as atomic energy, fuel cells, thermoelectricity, and solar energy. But it is unlikely that these sources of energy will replace fossil fuels in the foreseeable future. The adoption of these new energy sources will maintain the use of fossil fuels at the present rate and prevent further deterioration of the quality of air over our metropolitan areas. However, the use of atomic energy is not without its own problems.

Major efforts in the control of air pollution have been aimed at minimizing the total quantity of certain pollutants being discharged to the atmosphere. More recent control efforts recognize the importance of the time and spatial distribution of materials being discharged in addition to the quantity. Attention devoted to minimizing the discharge of pollutants at the sources will continue in the future because our ability to manipulate the atmosphere is nil. For example, it has been estimated that the energy required to maintain a nominal artificial 9 mph wind over the Los Angeles basin would approximate the energy-producing capacity of 12 Hoover Dams.

One of the contributing factors to the complexity of the air-pollution problem is the omission or minimization of the importance of "effluent quality" as a design criterion for the processes and devices which are necessary for sustaining the activities of our society. The typical pattern is to optimize the energy transformation process and consider the effluent quality as a separate problem. This may be entirely necessary, at least until our technological capabilities allow us the choice of a variety of feasible and economical processes for any given task. For example, the auto-exhaust problem will become less important as some other energy transformation process such as fuel cells or batteries become practical. Also as considerations other than economic ones become more prominent, industrial processes will incorporate the quality of the effluent as a process design criterion.

There is some evidence that more stringent control and control of a totally different character than practiced today may be necessary as our medical knowledge becomes more specific. For example, the role of combustion products in the elicitation of allergic responses in individuals is only vaguely understood. In the more distant future, the consequences of a global increase in the concentrations of substances such as carbon dioxide and sulfur compounds must be evaluated.

Succinctly, changes in the characteristics of the air-pollution problem are likely to emanate from the dominant use of new energy transformation processes, vastly improved technological capabilities which will allow a certain degree of manipulating of the atmosphere and the incorporation of effluent quality as a process design criteron, improved medical knowledge, and the acceptance of considerations other than economic criteria.

The Toxic Environment

Theron G. Randolph

Swedish Covenant Hospital
Chicago, Illinois

Lutheran General Hospital
Park Ridge, Illinois

The greatest single change in man's surroundings during the past century has been the ever-increasing chemicalization of his environment. Although this process started off with coal-burning home-heating units, kerosene lamps, and fuel-oil- and gas-burning household utilities, it remained for the modern petrochemical revolution to stamp the imprint of the chemical environment on the present generation. In addition to automotive products, the chemical environment also includes the widespread use of solvents, synthetic drugs, dyes, plastics, and fabrics, as well as detergents, pesticides, and other materials. Unparalleled chemical contamination of air, food, and water has resulted [1].

As judged by this ever-increasing rate of chemicalization of the environment in recent years, the habitat is becoming more and more menacing. It is already hostile to the point of impaired health and productivity for an expanding sector of the population.

Many questions arise in a discussion of this subject. For instance, how does the body react to lesser exposures of substances known to be toxic in greater concentrations? What are the distinctions between reactions on the basis of susceptibility (allergy) and toxicity? Are individuals capable of adapting to this relatively new chemical environment? Are we protecting ourselves adequately against our present chemical surroundings? Are our techniques of testing the safety of new chemical compounds sound? Do present practices increase the susceptibility of individuals, thereby enhancing the impact of the chemical environment, and threaten their future health and behavior?

It is well known that man possesses an amazing ability to adapt to changes in his intake and surroundings, provided such variations occur slowly. But ecologically speaking, the chemical environment has been an explosive development. Since a large sector of the population is already believed to be maladapted to various environmental chemicals, what may be expected of a future which promises greater concentrations of present materials plus many additional ones?

Descriptive features of these maladapted responses will be outlined. Interrelationships between reactions on the basis of individual susceptibility and toxicity will be emphasized to serve as a background for proposed changes in testing and regulatory functions.

The development of adapted responses to chemically-derived materials is exceedingly subtle and relatively unappreciated either by victims of this process or their physicians. When such relationships are demonstrated, the startled individual often asks: "Why did this just suddenly start to bother me?" Actually, the process had been building up for some time as a result of the hammering impact of frequently-repeated small exposures.

When a previously well person is first exposed to this type of chemical impingement, there is usually no obvious reaction. A particular agent, not being suspected, often results in an "I can take it" or "That does not bother me" attitude. Neither is adaptation to continued regular exposures in the presence of individual susceptibility apt to be recognized.

Instead of feeling worse, a person who is specifically adapted tends to be relatively stimulated immediately following each reexposure. These oft-repeated or constant-sized doses not only tend to maintain a relatively stimulated state, but some persons actually resort to such items as often as necessary to remain "normal." Most commonly, however, a specifically adapted individual merely carries on with his accustomed routines, oblivious of any type of response. The housewife reacting to the fumes of her gas kitchen range rarely ever suspects them.

But sooner or later, specific adaptation tapers off. The impact of exposure, enhanced by increasing individual susceptibility, gradually breaks through bodily defenses. Postexposure "pickups" then become relatively less beneficial and shorter. Delayed "hangovers" are manifested as increasingly chronic physical and mental illnesses [2]. Localized effects such as conjunctivitis, nasal disturbances, frequent colds, mild gastrointestinal distress, itching, or other lesser regional symptoms are first apparent. More advanced localized responses include asthma, hives, and eczema and musculoskeletal, gastrointestinal, genitourinary, and neurological manifestations. Closely related are general effects such as chronic fatigue, headache, and impairment of higher integrated cerebral functions. The latter also include cutbacks in humor, initiative, and ambition, as well as forgetfulness, incomprehension in reading, inability to make decisions, and faulty concentration. Confusion, depression, and other psychotic behavior collectively referred to as ecologic mental illness may eventuate [2].

Such maladapted stages of chronic reactions may persist for many months or years, as long as accustomed routines are continued. Indeed, life tends to become something to be endured in sick boredom, instead of a challenging and exhilarating experience. But if the responsible agent or agents are avoided sufficiently long to permit "hangovers" to subside, recovery recurs. Reexposure to such previously avoided incitants then induces acute immediate reactions. This course of events—specific avoidance followed by reexposure—changes chronic illness of obscure causation to acute illness and demonstrates its inciting and perpetuating environmental causes.

Susceptibility and maladaptation to multiple chemical and other environmental exposures are the rule, due to the tendency for cross-reactions to occur to materials of common genesis. Therefore, the simultaneous avoidance of probable, potential, and suspected agents prior to testing singly is most rewarding. This program, called comprehensive environmental control [3], has been carried out in an ecologic unit [4]. The effects of single reexposures are first observed in the hospital setting, subsequently as subjects are returned to their homes and to their work.

With the application of this technique, the significance of chemical exposures as causes of chronic illnesses in the average householder gradually emerged [1]. One by one, the effects of synthetic drugs, pesticides, and certain other chemical additives and contaminants of the diet and indoor chemical air contamination were incriminated. Of the latter, fumes from gas utilities, sponge rubber padding, plastics, and airborne insecticides were demonstrated to be most troublesome.

Although any person may adapt for a time and then maladapt and sicken as described, this exposure-reaction pattern occurs more readily in selected individuals. Such a selective susceptibility has long been noted in occupational situations, even though dealing with allegedly nontoxic levels of exposure. Greater quantities, known to induce reactions in most organisms, are usually referred to as toxic or poisonous exposures.

The important point to emphasize is that this exposure—reaction pattern, as it now occurs in the householder, is appearing as an ever-increasing factor in chronic illness. Regular repetition of minimal chemical exposures induces adapted responses readily confused with apparent tolerance. Relatively larger or more rapidly absorbed doses to which one had been adapted in smaller quantities, might break through and precipitate acute immediate reactions. When such exposures occur by chance, they are apt to be confused with those of acute toxicity. But when these governing circumstances are controlled, this technique of avoidance prior to reexposure may be used deliberately for diagnostic purposes.

In other words, the reaction of a person to a given amount of a specific substance may be interpreted meaningfully only when the individual's degree of specific susceptibility and his stage of adaptation to that and related exposures are known. From this standpoint, a sliding-scale relationship exists between reactions on the basis of individual susceptibility and those attributable to toxicity, rather than the rigid quantitative distinctions alleged to exist [5].

It is evident that many past decisions regarding the safety of environmental chemical exposures have not been made on an ecologically sound basis. Otherwise, there would not have been a continued increase in the number of persons becoming susceptible to and made ill from the impingement of various "accepted" chemical environmental exposures.

Since a working knowledge of the stages of adaptation permits either an accidental or an intentional bias in the interpretation of long-term chronic toxicity studies, and since such tests involve the interests of both the manufacturer and the public, the manufacturer who hopes to profit from the sale of an item should not be given sole responsibility for conducting tests as to its safety. Although the public's interest in these matters is said to be protected by governmental regulatory agencies, actual protection varies from being reasonably good in the case of new drugs to exceedingly poor in respect to airborne insecticides, gas-fired cooking and heating devices, and janitorial supplies.

If man is to regain his health and productive capacity with any hope of retaining them as he moves farther into the chemical age, much greater emphasis must be placed on the ecologic aspects of human existence. Human ecology, a concept which regards the individual as an integrated biologic unit in constant changing relationship to his intake and surroundings, must be recognized and explored. Its dynamic approach opposes current relatively static analytical approaches. The latter includes the current medical emphasis on bodily mechanisms of disease and the dominant toxicologic approach of studying the environment by means of an analysis of fractions of it.

More specifically, we need to incorporate the principles of human ecology and specific adaptation into testing techniques to evaluate new chemical agents and to reevaluate certain presently accepted applications.

In addition to these basic changes in techniques, we also need to elevate to a professional status those responsible for testing and approving chemical applications, freeing them more completely from production ties and associated economic pressures. As a society, we do not accept statements of corporations as to the status of their finances; independent professionals, i.e., certified public accountants, render this service. In view of the public stake in the matter of controlling the chemical environment, we should not accept the producer's statement as to the safety of his product; independent professionals, i.e., "certified public testers," should render this service.

If this appraisal of the health hazards associated with the chemical environment [1] and this critique of testing techniques [5] are correct, no time should be lost in establishing the educational facilities, licensure provisions, and ethical standards of so-called "eco-testers."

SUMMARY

1. The chemical environment, including gas, oil, coal, their combustion products and derivatives, is to be regarded as a composite environmental exposure, in view of the tendency for reactions to spread to synthetic materials of common genesis.

2. A sliding-scale relationship between reactions on the basis of individual susceptibility and toxicity exists, instead of the rigid quantitative distinctions presently alleged.

3. The impact of lesser chemical exposures known to be toxic in greater concentrations is magnified in susceptible persons and manifests as a wide range of chronic physical and mental illnesses.

4. The resulting health problem is best approached ecologically in keeping with the stages of specific adaptation. Medically, this means greater emphasis on demonstrating environmental incitants of illness. Toxicologically, this entails improved techniques to establish the safety of chemical exposures. Recognition of the independent professional status of the testers is also needed.

REFERENCES

1. Randolph, T. G., Human Ecology and Susceptibility to the Chemical Environment, Thomas, Springfield, Illinois, 1962.
2. Randolph, T. G., Ecologic Mental Illness—Levels of Central Nervous System Reactions, Proc. Third World Congress of Psychiarty, Vol. I. Univ. of Toronto Press, Montreal, pp. 379-384, 1961.
3. Randolph, T. G., "Ecologic orientation in medicine; Comprehensive environmental control. in diagnosis and thearpy," Ann. Allerg Vol. 23, pp. 7-22, 1965.
4. Randolph, T. G., "The ecologic unit," Hospital Management, Vol. 97, 00. 45-47 (March), Vol. 97, pp. 46-48 (April), 1964.
5. Randolph, T. G., "Clinical manifestations of individual susceptibility to insecticides and related materials," Industrial Medicine and Surgery, Vol. 34, pp. 134-142, 1965.

Urban Area Environmental Spectrum

Frederick T. Aschman

Barton-Aschman Associates
Chicago, Illinois

My brief comments on the future of urban areas will be framed in terms of the relationship of technology and public policy development.

The barest characteristic of the future city, and of course the easiest to predict, is bigness. For some 7000 years there have been cities of different sizes, types, and functions, but it has only been during the past 200 years that urban growth has accelerated to become the striking phenomenon it is today. A combination of industrial and agricultural revolutions has taken us from a nation in which in 1790 there were only 24 communities with more than 3500 people to a point where we can readily foresee a quarter of a billion people living in great metropolitan complexes in the year 2000. Estimates place the urban growth of the nation in the 50-year period from 1950 to 2000 at an increase of urban land use of some 20 million acres, from about 9 million acres of urban land today to nearly 30 million only decades hence. Nor does there seem to be any doubt of the general geographic forms of these future concentrations. The 600-mile-long eastern seaboard represents a pattern which is repeated in the Detroit—Chicago corridor and in the area from Toledo and Cleveland to Pittsburgh. On a smaller scale, we see the same kind of linear concentrations along the old route of the Erie Canal, in North Carolina's research triangle, and even in downstate Illinois. Thus, the town designers' dreams of rigidly controlled geometrics of city growth, of well-ordered rings of satellites, or of broad-acre cities have been dreams and nothing more. Neither public policy nor technological innovation is likely to overcome the momentum of today's tumultuous tendencies. The real questions in urban planning today are instead those dealing with the quality of physical environment and the compatibility of social structure. The challenge is how to achieve, within a predictable regional physical form, such objectives as those set forth in Chicago's recent official planning statement, "to improve the living environment, to strengthen the economy, and to enlarge human opportunities."

In seeking to deal with these questions today, urban planners are finding themselves less concerned with the familiar master plan than they are with a clear expression of public policy and the projection of definite programs of action.

In this process, technological capacity is assumed to be virtually unlimited. We know that there will be some sort of new form of transportation developed if it is demanded. We rely unhesitatingly upon the potential for controlling pollution or spoilage of our resources, and upon the potential for developing new designs and methods of construction of our structures and systems. Indeed, in this version of the planning process, technological innovation is simply to be expected, not predicted. It is seen as an essential device to be utilized as the need is established to advance the goals and directions set by public policy. The urgent planning gap is not then one to be filled by technology. Rather, our most urgent need is to improve our capacity for human decision in determining how to use our technological resources to achieve a better environment. It is only after public policy has been determined that we can bring the full force of technological innovation to bear in solving environmental problems of gigantic urban growth. If this is so and our technological capability can be assumed, then the question of our ability to formulate adequate public policy becomes the most critical question in urban planning today. I would suggest two important requirements, one of interest to us generally as citizens, and the other of interest in terms of the subject of this discussion.

First, there is the question of governmental capacity for policy development in areas of broad metropolitan significance. In our 200-odd metropolitan areas, there are some 17,000 local governments. Usually only the central city has any degree of comprehensive approach to future growth, and this in virtual collaboration with the federal government. In far too many instances, this leaves the vast suburban areas of the present and future without any real planning endeavor. The states, rather than any new monolithic forms of metropolitan government, offer perhaps the only practical approach to metropolitan planning. Yet, if the states are the most logical agency, they are also the most laggard. In any event, a more effective measure of metropolitan policy-making capacity is critical to our future.

Second, and perhaps most basic and interesting to us today, is the need for greater intimacy between technological expertise and political leadership at all levels of government. What is required is for the scientist to communicate to the political leader a full understanding of the potential of technological capability in the achievement of a better urban life. The comprehensive planning agencies that are now accepted arms of government are the logical channel for this communication. In this way, technological concepts must be advanced in sociological and hence political terms. The image of the new environment must be conveyed to the political leader who can solicit the public support that is essential to adoption of effective public policy. It is in this way that we'll secure the expressions of public policy that are a requisite to employment of technological capability to its fullest extent in environmental improvement.

Discussion

QUESTION: I'd like to direct a question to Dr. Bond, if I may, concerning psychological effects in this synthetic atmosphere of 80% helium. Is the frequency of voice raised as it is when breathing pure helium for a short time?

DR. BOND: This is a very complex subject, and I'm not qualified to discuss all aspects of it, but in general the frequencies go up almost an octave; some of the formants are simply lost, and the result is generally the effect of a couple of chipmunks in a good, hot fight. Psychologically, this can produce a very real problem inasmuch as when men are placed together in a hostile environment and are deprived of the means of intercommunication, all hell can break loose, you might say.

QUESTION FROM THE FLOOR: I'd like to direct a question to Dr. Cember. Do you have any comment on the possible genetic effects of environmental agents other than radiation, such as drugs that we use, or other materials with which we come into contact?

DR. CEMBER: I know that there are a number of other environmental agents that cause genetic effects. I don't recall their names now, but I do know the list is relatively long and that people working in this field believe that hazards from these agents are real. One name that comes to mind of a man working in this field is that of a man named Sobols. I believe, at the University of Leyden in Holland. He has shown that there are other environmental agents that will produce essentially the same effects as radiation, and one of these that comes to mind is caffeine in coffee. I believe that if you make the most pessimistic calculations in the case of caffeine, the results are of the same order of magnitude as the results of similarly pessimistic calculations in regard to radiation. There are other environmental agents that we have to consider, but I can't name them.

QUESTION: Dr. Randolph, I believe that you advocate that all possibly toxic materials should be certified as to safety. We already have licensing laws of people qualified to make this certification, namely, physicians. Needless to say, a physician might need some graduate studies and should not automatically go into this field without preparing himself, but I think the law is already set up to protect the public. What would be needed is the willingness of individual companies to cooperate. Perhaps, they would have to be forced by law to submit to certification. Of course, if a physician certified something as safe and it wasn't, he would have the professional responsibility for it and face loss of license and possibly jail. So, I think the public could be protected if it wanted to be protected.

DR. RANDOLPH: Obviously, I cannot outline the details of how this should be changed, but suffice it to say that there has been a tendency in the past for our toxicologists to regard many exposures to be safe in small amounts which are notoriously toxic in larger quantities. I think that this is an unreal distinction. Many people subjected to such day in, day out reexposures as fumes from gas ranges and traffic exhausts become highly susceptible and chronically ill. Yet the amount of these exposures is said to be small. One cannot set a rigid threshold dosage, above which is "toxic" and below which is "safe". Individual susceptibility on the part of many people makes this a variable distinction. Indeed, individual susceptibility seems to be the crux of this problem. I can see only increasing illness in continuing the present assumption that lesser chemical environmental exposures are innocuous.

KUEHNER: I want to make a comment on this subject. I divide the future environment into two groups, in keeping with Dr. Randolph's approach, the indoor and the outdoor environment. Dr. Randolph has dramatized the accumulation indoors. In terms of your air-conditioning men, where a hundred years ago we had infiltration at the rate of 8 to 10 to 20 times an hour, currently we deal with two air changes per hour indoors. Now we're getting to tighter houses, which means this accumulation becomes more serious. Would an air-conditioning expert, perhaps Mr. Boswell, like to comment in what direction we are going in the way of air changes inside, i.e., infiltration air changes. Is this problem going to become more serious?

COMMENT (CY BOSWELL): I am no expert on infiltration, but it is true that houses are becoming much tighter. We spoke earlier, at the previous panel discussion, about the need for humidification in the wintertime. There are some houses that are now constructed so tightly that there is need for dehumidification in the winter rather than humidification, so that this increase—decrease in infiltration is certainly a factor in the construction of present-day homes and buildings.

QUESTION: Mr. Aschman, we were talking about the pollutants in the air and controlling them at the source. The conclusion was that you had to control them at the source. There was nothing you could really do except to control the source. In Los Angeles a blow-by system on exhausts has been introduced. I think it costs something like $40 or $50 a year to service, and there is resistance to it. Now, you say that the technologists must get the ear of the politician in order to see that the environment is fixed. We were told by Dr. Terry that President Johnson's Great Society has to exist in a better environment for man, so conditions do look ripe. It looks like a good atmosphere to have this carried on, but do you think it is going to work? Is it going to fly?

ASCHMAN: I suspect that if the knowledge of environmental control that exists in this room today could be adequately transmitted into the political arena, into the arena that determines regulations, we would probably make more progress in regulation overnight than we have made in the last 25 years, because actually our whole method of regulating would be well established by performance standards and zoning ordinances. In the area of odor, for example, we might simply say that an industry may not locate within so many feet of a residential area if one can smell it. The real point is that I think we do not have an adequate channel of communication between the technological arena and the political arena. It is just that simple. There is a great deal that we could do within our present governmental framework, within our present concepts of property, law, and so on, if we simply knew more, and I suspect that those of us who draft these regulations and who help to formulate the public policy just have a mere smattering of the knowledge that actually exists. I think that we could make tremendous strides in a very short period of time, if we could improve that channel of communication.

QUESTION: But basically, would the public accept this? A man wants his car. We know that from an architectural standpoint, the city planning standpoint, and the transit standpoint an individual all by himself inside a 20-ft-long car is the biggest problem in the world. Not only is he a problem architecturally, but he also pollutes the air as he goes along. But this car is his last hold on his individual freedom, I think, and we just can't stop him from using it. Now, what are we going to do about the man at the bottom?

ASCHMAN: This may sound naive, but over the past 25 years I've had a rather intimate relationship with this kind of a problem, with "what do I think is good" versus "what the public thinks is good." I would say very firmly that in the vast majority of the type of planning proposals we advance and lose, we are either wrong or we haven't made a case. In other words, I have great faith in the fact that the public will accept things if the proper image is projected in terms of public benefit. Winston Churchill is supposed to have said of one of the famous town planners of Britain, whom he didn't like, as the fellow walked by: "There by the grace of God, goes God." I think we have a tendency to try and project public policy on the basis of our simple statement that we think it is good. But if the necessity for these policies is demonstrated in terms of public welfare, the vast majority of proposals are accepted. This is where political leadership comes into play. But we must translate the technological image into the image of public benefit in such a way that the political leader can in turn gain the public support for it. We can't just stand up as technicians or scientists or planners and say we ought to have this because it is good. But the political leader is really a key to this. I have a great deal of confidence that we can go further than you think right now.

KUEHNER: Dr. Collier, would you like to extemporize on this subject of communication? I know you've been concerned with it.

DR. DON COLLIER: Well, of course, it is always a problem between any two bodies of people who have different backgrounds, and I think more of these conferences will help bridge that gap. I am delighted to see you working at it.

A.J. HAAGEN-SMIT: I want to take issue with the previous speaker when he states that my car is the only freedom left. We in Los Angeles are much aware that freedom to move is severely curtailed in the chronic traffic-tie-ups at 8 o'clock in the morning and 5 o'clock at night. I would also like to reply to the suggestion that the standards used in Los Angeles are stringent. There is a fundamental difference between standards for industry and those for a community. In industry, we have a selected group of people; in a community there are people of all ages and of various degrees of sensitivity and health. One has to be very careful in judging statistical data. When statisticians tell us that there are only a few percent of days with eye irritation or that there is only a few percent chance that one dies, it is somewhat disturbing to think that you might belong to the statistically insignificant unfortunate few. Actually, these few may be very large numbers in our densely populated urban areas. In Los Angeles, we adopted a community health standard for ozone or oxidant in the air which should not exceed 0.15 ppm. At this level, 95% of the people are not bothered, but conversely it is also true that 5% are bothered. Five percent of 7 million people is 350,000 people, enough to elect or reject a president of the United States, let alone a supervisor of a county.

Such calculation shows that standards are most often not too severe. On the contrary, we have to reduce the levels gradually when engineering methods are developed further in order to cope with the increase in population. This lowering of standards is contemplated for auto exhaust in 1970. The new 1966 cars in Los Angeles will have corrected emission from the crankcase, as well as from the exhaust. This is important progress, and already the industry is developing better control at a more reasonable price to cope with the situation many years ahead.

The Los Angeles Chamber of Commerce will be happy to know that there is still hope for Los Angeles.

QUON: I am in reasonably full agreement with Dr. Haagen-Smit. I did not mean to convey the impression that the present standards are inadequate, but merely that they are incomplete. With regard to the automobile exhaust control, there is a conflicting problem because one is interested in oxidizing hydrocarbons, which is one of the components of the photochemical smog, and one is also interested in reducing the oxides of nitrogen.

EDWARD HERMANN: I would like to address my comment to Dr. Quon, and will shift from the oxides and hydrocarbons to particulates that have to do with air pollution. We see some criteria for stack discharges based upon the emission concentration at the top of the stack and others based on concentrations downwind. I would like to have some comment on both types of criteria that have been developed.

QUON: I think that the emission-type standards are somewhat easier to apply and enforce, although one is concerned with the concentration of particulates at ground level. In terms of specifying a standard for the quality of air, it is probably more direct to specify the downwind concentrations. The downwind concentration is made up of emissions from several sources, and the allocating of responsibilities and partitioning of the concentrations to the various sources are difficult. Both standards are being used, and they do help to regulate the discharge of materials into the atmosphere. As to which is more preferable, I would think that, under the condition of implementation, it certainly is easier to have the emission standards.

COMMENT (HERMANN): In connection with the Chicago ordinance, it appears that the emission provision in the code of 1958, I believe, is 0.35 grain per cubic foot maximum, with some allowance for additional stack height. This is about 800 milligrams per cubic meter, if my arithmetic is correct. Yet we see a possible problem here. Most woodburning fireplaces in a home will probably put out more than this because the ash-to-air ratio going up the stack is a function of the fuel-to-air ratio involved. A man with a quite small stack, say a 2-in.-diameter smokestack, if you can imagine it, could be forbidden to discharge it, whereas someone discharging a thousand times more, say a 10-ft-diameter smokestack, is still in business if he can cut down the ash going up by blowing steam and boiler gas up the stack.

QUON: I think the criterion is reduced to a standard CO_2 concentration. The issues you raise certainly are complex. Both the large pollutors and individuals—indirectly, through their activities—contribute to atmospheric pollution. For example, in the San Francisco area it is estimated that the industrial pollution is roughly half of the total and that domestic sources such as automobiles and individual activities contribute roughly half.

JOHN BIORK: I was wondering if any of the panelists had any experience with either health or environmental effects registered by persons in the extremely clean rooms that are now being used in industry. This would seem to be the perfect environmental chamber, with people working in it from time to time, making bearings or whatever they're doing in these extremely clean rooms. Have there been any studies made on the environmental or health aspects of these people? I know they're not in them all the time, but is there a reduction in disease or virus effects, colds, or the common ordinary health hazards encountered by people not working in this type of environment in similar industries?

KUEHNER: Any comments? How about Dr. Behnke? I think this is sort of like the old submarine, where you clean the air by adhesion to the oils along the surfaces.

BEHNKE: It is very interesting. Dr. Bond has made a study of this too. In the submarines, after the first week or so a virus will go around and about three-fourths of the individuals will have colds. Then, after that, for a period of weeks they are well. For the rest of the cruise, perhaps three months, there will be no colds. There will be no respiratory infections. The individuals apparently have immunity, and they're free from colds. Later, of course, back on shore, with new contacts, they again acquire whatever goes around in the air.

Now, may I ask Mr. Aschman a question? For a number of years I've been at the radiological defense laboratory and I've broken my back on the problem of underground shelters. I think that I should have gone under water, like Captain Bond. But, we can't do that, and I would like to ask if any consideration is being given in urban planning to anything really going underground? Not necessarily homes—people won't do that—but I'm talking about office buildings. I'm talking about underground garages and all that sort of thing with parks on top.

ASCHMAN: I know of no tendency anywhere to do this for protective reasons. The only actual underground developments that we get, such as underground garages, for example, are motivated by economy of land and nothing more. I think that in this matter of developing the public image it should be recognized that in many cases, the public doesn't really react as strongly to things that are devised to protect them as they do to those things which would bring them some concrete benefit. I think we have to understand the basic political maxim in our society today is that control over such matters of public policy is vested largely in the upper middle-income group and the low-income group. The upper-income group develops a civic consciousness in this regard, and the lower-income group has nothing to lose. It is this coalition which is aimed mainly at positive benefits rather than at matters in the protective arena, which may be very, very important, but a response to which is much more difficult to achieve.

AL NEWTON: My question is addressed to Mr. Aschman. If I understood you right, I believe you indicated that you felt organized growth of patterns, perhaps of cities, is hard to attain. I just wonder if we have worldwide communication on such things. It seems to me that some cities, take London as an example, have been able to establish some very good patterns of growing satellite communities and perhaps offer different incentives for people to move into them. Does this establish some law or some precedent that might help us here?

ASCHMAN: I think your point is well taken. The example isn't so

good because in England the strong tendency is to go away from a
satellite city for social reasons that they really haven't achieved,
but that is rather beside the point.   I think I should have qualified
my remarks to limit them precisely to the United States, because
we get the concept of predetermined geometric forms in cities
and countries where property law and the view of property rights
versus human rights are considerably different from ours.   I think
most of us here do not see any real opportunity (not that it would be
particularly desirable, because the price might be too high even in
these other terms) to so thoroughly regulate growth that we could
achieve these theoretically perfect geometric patterns.   Again, the
price would simply be what most of us think to be too high in terms
of the institution of property.   I think we can have some influence
over the form of cities, in terms of encouraging a particular form,
by the type of public services and development of public systems
that would encourage these particular directions.   But I think in
most cases where you see a city developing in other countries to a
predetermined, definite geometric pattern, it is either because
the dynamics of growth are not what our dynamics are or because
they have an authoritarian control over that geometric pattern which
we do not have, and which presumably we do not want.

RICHARD ROSENBERG: Mr. Aschman, with your experience in the
Chicago area, based on the technological information available and
the results that have been achieved by the Northeastern Illinois
Metropolitan Area Planning Commission, do you think we will
achieve control over water and air pollution in this area before
catastrophe strikes?   Do you really think there will be cooperation
before the real point of catastrophe

ASCHMAN: I frankly don't know.   I think present conditions entitle
us to some skepticism in this area.   The fact is that in terms of the
Northeastern Illinois Metropolitan Area Planning Commission, the
state of Illinois sees fit to appropriate only $50,000 a year to this
six-county planning operation.   By contrast, the City of Chicago
has a budget of over a million dollars a year for exactly the same
purposes within the city, and then, in addition, the city contributes
$50,000 a year to the Metropolitan Planning Commission.   What this
really means, in my opinion, is that the arm of government which
should be most interested and concerned with this problem is
really not very interested, and that is the State of Illinois.   Now,
we may see this changed in view of the one-man, one-vote decision
of the Supreme Court, where we may get a state government which
is much more oriented toward and aware of metropolitan problems.
If we do, I don't know to what extent or how far we will go, but at this

point, the Metropolitan Area Planning Commission is relying simply on the hand-to-mouth feeding from voluntary contributions by miscellaneous governments, plus a rather substantial amount of citizen participation. Dean Gotaas, for example, heads the very important technical advisory committee that we couldn't buy for all the money we have in our budget. But there is not really at this point any real evidence of concern at the very level of government where the concern ought to be.

DON McPHEE: Dr. Bond, you mentioned that the problem of communication in your artificial atmosphere had been solved. Could you tell us what the solution was?

BOND: The solution was a quick and dirty one, resulting in a breadboard model which in effect chopped off all frequencies below 3000 and all frequencies above 5000 and heterodyned the remainder. It was declared that this would not work. Obviously, this is cutting into the normal conversational level. However, it did work, and it resulted in a mechanical type of speech which was intelligible. It has about 80% intelligibility. We are looking forward to considerable improvement in this design, but this is a problem which was solved. The thermal problem has not been solved. When we put the men out in the water we are going to be in very severe trouble.

I would also like to add a comment to Dr. Behnke's remarks about our nuclear submarines. We have identified over 200 potentially toxic substances in the atmosphere of our submarines. This, in spite of a very, very careful program to prevent the introduction of dangerous substances into this atmosphere. These people, as you appreciate, are breathing this mixture containing potentially toxic substances for continuous periods up to 90 days. We have, only this year, received consideration to go ahead with a program of actively observing the people who have been so exposed over the next 20 years of their lives, so that we might come up with some concept of whether we are damaging these men or not. This is the present state of the art. It has taken us 15 years of arguing to get permission to do this.

Man and His Environment:
Where Are We? Where Are We Going?

P. B. Gordon

Wolff & Munier, Inc.
New York, New York

INTRODUCTION

When Professor Jennings called me in October to test my acceptance for this talk, I asked him about the proposed program. I tried to sell him on having a more scientifically oriented speaker for this assignment, but he advised that the planning committee wanted someone from industry. My remarks therefore are those of an engineer-contractor who has spent most of his working years in one environmental area — that of installing systems to provide "thermal comfort" for buildings.

In the process of preparing my talk, I drew on several different kinds of experience: as an engineer-contractor for HVAC work, as a member of technical committees and as an officer of the ASHRAE, as a member of the Board and an officer of the BRI (concerned with total environment), and as a trustee and officer of the John B. Pierce Foundation (concerned with basic research in thermal and sanitary environment).

I did intend to review and highlight the program of the past two days. This I was asked to include in my remarks but, to save time and because you have heard the several good presentations on thermal, acoustic, and light environment, the social and medical aspects of toxics, air pollution, and odors, and the pressures of population on urban planning, I will simply say that all the presentations were illuminating and interesting, and the sum total effectively highlights the general theme: "How Man Interacts with His Environment."

Instead, I want to talk on two aspects of the total environment problem. First, because my experience has been in the area of "thermal environment," I want to review the present state of the art, where we are, why more work is necessary. This I intend to do without duplicating the excellent presentations in specialized areas by Dr. Lee, Capt. Behnke, and Mr. Viessman. And to conclude, I want to briefly suggest some ideas in the general area of "total environment."

THERMAL ENVIRONMENT

Thermal Theory of Ventilation and Modern Air Conditioning

In recent years, cities and states have been revising their building codes to recognize modern air-conditioning practice as an engineering technology, capable of controlling the indoor thermal environment. To illustrate, during 1956-1959, the City of New York adopted air-conditioning rules that accepted modifications of the 1938 "Ventilation Rules," still part of the Code. These ventilation rules and the "ventilation index" relate cubic feet per minute of outside air per square foot of occupied floor area to the number of occupants, the cube of the space, and the area of operable windows. The resulting index determines that the cfm/ft^2 of supply outdoor air shall be zero, 0.33, 1.0, 1.5, 2.0, 2.5, or 3.0 with corresponding values for exhaust quantities.

The outdoor air ventilation requirements of these rules were intended to remove the heat and water vapor released by occupants from the occupied spaces, rather than to remove objectionable odors or freshen the air.

The "thermal theory of ventilation," that ventilation for occupied spaces was required to protect against excessive rise in temperature and relative humidity, rather than to control the chemical makeup of the indoor air, was rather well accepted by 1900-1910, based on the work of German, English, and American investigators. In today's urban atmosphere, with the problems of industrial pollution being so pertinent, I am not so certain that the indoor environment is improved by introducing excessive quantities of outdoor air, which is not always "fresh air." However, since man must live and work indoors for long periods of time, air conditioning can provide a composite atmosphere indoors that is capable of controlling, within acceptable limits, the dry-bulb temperature, moisture, air motion, and, when necessary, the surrounding surface temperatures and the cleanliness of the air. This will provide thermal comfort for most of the occupants and provide that feeling of air freshness so much desired by most people. In other words, it will provide a satisfactory thermal environment.

It is interesting to note that the final acceptance of the "thermal theory" of ventilation, during the first 10 years of this century coincided during the same 10-year period, with the first approaches to the problem of total air conditioning. This requires separate control to some degree of accuracy of the dry-bulb and dewpoint temperatures of the air supplied to the occupied spaces. This was the same time that the mathematics and thermodynamics of controlling the psychrometric properties of air were being firmed up. The installations of that period were few and, generally, of a special nature: systems for industrial, munitions, and chemical plants, with but few specifically for human comfort.

During the 1920's, the big expansion was for human comfort, principally for movie theaters. This work was done (the engineering, manufacturing, and installation) by three or four companies that fairly well cornered the market on know-how and equipment design. Society data on comfort requirements, heat release by occupants, engineering calculations, etc., had not yet found its way into the general literature (at least not until the late 1920's and early 1930's). During the 1930's, the groundwork was being laid for the tremendous expansion that took place during the post World War II period. New equipment and new approaches were being studied and developed. To name but a few: development of the freon refrigerants, direct expansion equipment, lighter and more compact equipment, finned surface cooling coils in lieu of air washers for coolers and dehumidifiers, and self-contained factory assembled air-conditioning units. In turn, this expansion meant new organizations were being set up for design, for contracting, and for manufacture. Owners were becoming aware of the value of comfort control to increase sales and to improve worker efficiency.

World War II was of great importance in this advance because of the large number of people that worked in the blackout plants, and the special military requirements for air conditioning. As we well know, the big push started in 1946, after the war. To use the New York City office building picture* to illustrate:

- During the 1925-1933 building boom, 30,000,000 ft^2 of office space was put into use, and very little of this was air conditioned. This included the major bank buildings, a large part of Rockefeller Center, the Empire State Building, the Chrysler Building, etc., etc.
- Since 1947 (1947-1964), 172 buildings equalling 62,700,000 ft^2 have been completed, and all air conditioned.
- To be completed in 1965, 3,700,000 ft^2 more, also all air conditioned.
- Already now projected for 1966-1970, 20,000,000 ft^2, also fully air conditioned.

What did this do to the 30,000,000 ft^2 of 1925-1933 buildings ? The conversion to air conditioning started in the early 1950's, and now substantially all have been converted to all-year conditioning. Many cities reached the 100% figure for air conditioning of all office

*The New York Times Real Estate Section, Jan. 10, 1965.

buildings many years ago, with Dallas and Houston as early examples, about ten years ago. And the office-building story applies in the same fashion to factories, hospitals, hotels, motels, and today residences and apartment houses. As more people live and work in thermally-controlled buildings, the more is the pressure for other buildings to do the same.

This development is a good example of the rather long time lag that often occurs between scientific knowledge and applying this knowledge to human needs. Two hundred years elapsed between the discovery of the microscope and microorganisms by the Dutch philos-opher–naturalist Leeuwenhoek and the application of these findings to the field of public health by Pasteur in the mid-19th century. A similar delay of more modern import is the lag between the research work of Goddard on jet propulsion and the final application in recent years to modern missilry. In air conditioning, this lag between scientific findings and their application to the handling of air and water vapor mixtures (psychrometry) was almost as long. Boyle started keeping a record of psychrometric observations at Oxford in 1666. In 1786, Saussure wrote about the wet-bulb thermometer and the need to whirl it to obtain a wet-bulb reading. Apjohn in 1835-1837 disclosed his thinking on the relationship between wet-bulb temperature and total heat. Regnault in 1845 wrote several scientific papers on psychrometry relating to dewpoint, vapor pressure, and relative humidity. Psychro-metric tables were published in London before 1850.

While there are many recorded examples of air cooling without humidity control that was installed before 1900, why were these scientific bits of information not put to real use before 1900-1910 or, more so, 1910-1920? Because the need did not exist, or was not recognized. In other words, there was no compelling need to demand the development. Air conditioning or the control of thermal environ-ment is not the only technology that developed rapidly when the need became important. As mentioned before, much of the technology of air conditioning, as it is practiced today, has developed since 1940, and, principally, because of two compelling needs, (1) the needs of World War II and (2) the needs exerted by the changes in building design. Again, using the modern office building to illustrate the changes in building design that are related to or result from the advances in air-conditioning technology, let me cite following:

- Air conditioning permits the use of floor areas that were hereto-fore considered submarginal, i.e., areas well away from windows are rented at full value, or by eliminating street noises and dust, makes the lower floors as acceptable as upper floors.
- Air conditioning permits the use of less expensive floor plan arrangements, i.e., there is less need for narrow wings or light

and ventilation courts. (Perhaps, in the story of "total—total" environment, both physical and aesthetic, this may not always be considered an advantage created by air conditioning, but more on this later.)

- Air conditioning permits higher occupancy per square foot of floor area.
- Air conditioning permits the higher lighting intensities now desired.
- Air conditioning provides for better housekeeping.
- Air conditioning provides for improved worker efficiency.

Standards of Thermal Comfort for Humans

What we call "thermal comfort" is really being in equilibrium with one's thermal environment, or, to use the definition of the Thermal Comfort Standards Committee, "that state of mind which expresses satisfaction with the thermal environment." While comfort or rather discomfort in an indoor environment is affected by climate, season, activity, and health, age and sex, clothing he or she wears, the most pertinent controllable factors are: air temperature and relative humidity, the surrounding surface temperatures, air motion, and the relative cleanliness of the air (this last not really thermal but included because of its close tie to true air conditioning). Much of the work to define standards of comfort (at least in this country) is the work of the American Society (ASHRAE), with great assists by many others at various research laboratories. The concepts and terms "comfort chart," "lines of equal comfort," "effective temperature," and "comfort zone" began to work their way into the Society literature in the early and mid-1920's.

Dr. E. Vernon Hill prepared a "Synthetic Air Chart" in 1920, which was one of the first attempts to catalogue and to emphasize the importance of the several physical and chemical properties of man's atmospheric environment. Dr. Hill also reported during a discussion of the 1923 Comfort Zone and Comfort Chart that the term "comfort zone" had been used in 1913 by Professor John Shepard of Chicago Teachers Normal College. The work of the Society Laboratory in environmental responses, which at first was in association with the U.S. Bureau of Mines at Pittsburgh, started in 1920 and continued there until 1943-1944, and later in the 1950's at Cleveland. Our knowledge today as regards man and his thermal environment results from these Society activities in this area, plus (and I want to emphasize the plus) the work of many others in other laboratories.

This work developed through the years in the following areas:

- Studies of the responses of humans to different environmental conditions, both normal and stressful, at rest or working.

- Studies of physiological reactions and change during these environmental reaction studies.
- Mathematical evaluations of heat loss for the resting and exercising man for combinations of air motion, temperature, surface temperature, and evaporation.
- Studies of the effects of clothing on human responses, and on physiological reactions.
- Studies on the total problem of temperature regulation in man and animals.
- Studies as to the efficiency of workers, learning capabilities, etc.
- Studies of the effects (both physiological and comfort vote) on the cardiac, arthritic, asthmatic, obese, and thyroid patients, and, as related by Capt. Behnke, the special environmental requirements for hospitals, premature infants, and hospital operating rooms.
- Development of the apparatus and instruments needed for progress in these areas, such as: radiometers for skin temperature and environmental radiation measurements, gradient calorimeters, ultrasensitive scales, skin blood flowmeters, and heat flow-meters.

This was important work and the contributions were meaningful in the development of the comfort requirement information that ultimately became available to all. This work progressed in many places in this country, and to list but a few as examples and not the entire story: The Society's Laboratory (Pittsburgh and Cleveland), John B. Pierce Foundation Laboratory, University of Illinois (and its School of Medicine), Cornell University Medical School (and the Russell Sage Laboratory), Harvard School of Public Health, University of Pittsburgh, Kansas State University, U. S. Bureau of Mines, and many others in this country and abroad.

I want to highlight a few points. While the Comfort Chart of today does not markedly differ from the Comfort Chart of 1925, in its application to modern design there are certain differences:

- In 1925, the winter comfort zone extended over effective temperatures from 62 to 69°F, with optimum comfort at an ET of 64°F.
- By 1930, the winter optimum comfort line had shifted from an ET of 64°F to 66°F.
- Today, the optimum winter comfort line has been raised to an ET of 67–68°F.
- The earlier shaded-area comfort zone (originally effective temperatures from 62 to 69°F) has now been removed, but, if we use the 80% group that votes comfortable, then the winter range is now effective temperatures from 65 to 70°F.

- Using the 20-40% relative humidity zone for winter, this would mean that these shifts for the optimum effective temperature have resulted in the following dry-bulb shifts:

    1925 — optimum at 69°F (DB) ± 2°.
    1930 — optimum at 72°F (DB) ± 2°.
    1965 — optimum at 74°F (DB) ± 2°.

- Similar changes have not been so pronounced for the summer conditions:

    In 1930, the optimum ET was 71°F; today it still is 71°F. At 50% relative humidity, the corresponding dry-bulb temperature is 76°F, but the zone of acceptable dry-bulb temperature has been narrowed.

- Today's chart has deleted the earlier shaded-area comfort zone. If we use the 80%-voting-comfortable figure, the ET range has been narrowed to 69 to 73°F from the earlier 66 to 75°F. Again, this has narrowed the acceptable dry-bulb range to 75 or 76° ± 2°.

- The 1963 Guide Chart now shows new emphasized almost vertical lines for comfort that are superimposed on the effective temperature lines to show that the dry-bulb temperature in the comfort area is more pertinent than the relative humidity.

The deliberations of the ASHRAE Standards Committee on Thermal Comfort have been completed and should be available shortly. This new standard is intended to specify environmental conditions to provide year-round thermal comfort for most people engaged in sedentary or near-sedentary activities, and wearing normal clothes. This new standard will replace the Code of Minimum Requirements (the last official Comfort Standard), established in 1938 and withdrawn about seven or eight years ago. This standard recommended a sliding ET scale and a sliding DB scale as the outdoor temperature shifted from 80 to 105°F. The Committee, in developing the new standard, has taken advantage of the most current information in comfort research, environmental physiology, and the "feedback" of experience from commercial practice. As to "feedback," I want to illustrate with three field experiences that vividly emphasized thinking other than what the summer comfort chart set forth at the time.

The first experience, in 1938, involved an installation of air conditioning for several radio broadcasting studios. The design required operating at 78°F (DB), with relative humidity not over 50%. Due to a deficiency in the cooling apparatus, during the first several months of summer operation, the dew-point temperature of supply air could not be depressed sufficiently to achieve this operating level. The air quantity and the dry-bulb temperature were low enough to reach 78°F (DB), but the relative humidity would climb to 60% and 65%. We all kept our fingers crossed during three months of operation, but received no discomfort votes from the large number of occupants. The deficiency in cooling coil performance was corrected to permit achieving the desired relative humity.

The second experience, in 1949, involved a research laboratory and assembly area for vacuum tubes. Because all the air was outdoor air, and the refrigeration load varied directly with the outdoor wet-bulb temperature, for the peak design day an operating condition was selected from the then comfort chart of 80°F (DB) and 40% RH (because 40% RH maximum was required for the vacuum tubes). Most of the several hundred occupants voted "uncomfortable" when temperatures reached 80°F (even at 42% RH). Complaints would begin as the dry-bulb temperature exceeded 76°F, and grew in intensity as 80°F was reached. Because of a refrigeration limitation, we could achieve either 75°F (DB) and 50% RH, or 80°F (DB) and 42% RH. 75°F was fine for comfort, 42% RH was fine for vacuum tubes. Comfort votes from several hundred occupants were recorded for the entire first summer and the result was the addition of refrigeration equipment to permit reaching 75°F (DB) and 40% RH. The system has been satisfactory since then.

The third experience, in 1950, involved the use of a sliding-scale dry-bulb temperature control for the interior areas of a large office building. The control was specified to shift dry-bulb temperature from 74°F to 80°F, as outdoor conditions changed from 75 to 95°F (DB). As soon as indoor dry-bulb temperature would go above 76-77°F, the complaints would start. A month or so later the sliding-scale control was abandoned, and the control was set at 75°F.

For at least 10-12 years, there has been a gradual shift away from "equal comfort" lines that indicate several combinations of dry-bulb temperature and relative humidity for equivalent summer comfort. It is now well recognized that dry-bulb temperature is the more pertinent value when in the zone of body thermal equilibrium, and the importance of the effect of relative humidity on comfort has been minimized in this zone. Laboratory tests and field information now show that, with dry-bulb temperature controlled between 73 and 77°F, any relative humidity between 25% and 60% is equally acceptable. In this 73-77°F (DB) zone, relative humidity within these limits (25% and 60%, and even 70%) has little effect on the comfort and well-being of seated or slightly active people. However, as the dry-bulb temperature goes above 78°F, the humidity effects become increasingly significant because evaporation then becomes a major avenue for heat loss from the body.

Need for Continued Research in This Field

While satisfactory standards for comfort are available, resulting from more than 40 years of laboratory investigation in this country, including the continuing feedback from field operating experience through these years, we still have this important problem: Why can these standards apply to only most of the people most of the time ? A large area of work yet to be done involves the "reasons for thermal discomfort."   To illustrate, I draw on another area of personal experience and want to use some of the work now under way at Pierce to highlight this general question: What makes for thermal discomfort

First, let me tell you about Pierce Laboratory and its background. The Foundation was established in 1924, pursuant to the dictates of Mr. Pierce in his will. The charter spelled out, "...to promote research, educational, technical, or scientific work in the general field of heating, ventilation, and sanitation, for the increase of knowledge to the end that the general hygiene and comfort of human beings and their habitations may be advanced. . . ."

In accord with this requirement, the trustees went forward along two paths: (1) in housing research during 1931-1952, first in New York City and later in New Jersey, and (2) in the setting up of a laboratory in New Haven in 1934 for psychophysiological research in human comfort. It is this second path I want to talk about.

During the period 1934-1952, a major research program in the field of human ecology was supported, resulting in over one hundred publications in scientific, technical, and professional journals. The principal areas studied were physiological and biophysical conditions of human heat exchange, partitional calorimetry, human and animal. adjustment to the thermal environment, temperature-regulation mechanisms in man, relation of comfort and thermal adjustment, effects of clothing on temperature adjustment, clothing regulation and fabric properties for protection against heat and cold, and heat-radiation exchange for various fabric properties.

During 1952-1961, the Laboratory concentrated mainly on heat-transfer studies, on heat-transfer fluids, and on a hydronic application — valance heating and cooling. In 1961, when Dr. Hardy became the Director, the Laboratory resumed its emphasis on fundamental research related directly to man and his environment. Specific areas of current interest are temperature regulation in man, thermal sensation and comfort, respiration and air sanitation, air pollution, and thermal radiation.

The past work of Pierce shows up in the literature of the Society, in the many papers in the Transactions, and in journals of physiology and public health. Important contributions toward our knowledge of man and thermal comfort include: (1) bringing into the comfort studies the concept of "operative temperature," which included the effect of radiation; (2) defining the zones of evaporative regulation, vasomotor regulation, and body cooling with the concept of "wetted" skin area; (3) the tool of "partitional calorimetry" that permits a mathematical evaluation of the total body heat-exchange mechanism with its surroundings; and (4) the introduction of the "clo" unit for insulation of clothing.

I believe the advancement of total knowledge for the general comfort of man in his habitations and as related to his thermal environment still requires further research, and more than any laboratory can encompass. Present emphasis at Pierce is in this area and the closely associated area of air sanitation. Some of the questions being explored are continuations of earlier work, some are new questions. For example: What are man's physiological responses to the thermal environment? What physiological responses contribute to comfort and well-being? How is the thermal environment defined and how can these environmental factors be related physiologically to the impression of thermal comfort and discomfort? Is a man uncomfortable because he sweats? Because he feels warm or cold? Because his heart beats fast or slowly? As yet, I am not sure we know why a person is made uncomfortable in the heat or cold. To evaluate the influence of sensations arising within the body, a comprehensive program is under way on the role of the hypothalamus in man's temperature regulation and his physiological response to cold and hot environments. A surgical technique has been developed for changing by thermal transfusion the hypothalamic temperature of the brain of an animal. Its metabolic responses (shivering, sweating, panting, appetite) and neural responses (brain waves) are observed simultaneously.

Man's physiological response to his thermal environment is also affected by acclimatization. Field studies of man and animals who live at extremes of heat and cold are being made in Norway, Australia, Africa, Asia, South America, and Alaska. The scientific literature is now very extensive on the biophysics of man's thermal balance when continuously exposed to various levels of heat and cold, with and without protective clothing. However, there is still much to be learned about man's thermal equilibrium, if such a state can ever exist, and man's corresponding sense of well-being and comfort. He is constantly changing his physical location and hence his thermal environment. The biophysical effects of a sudden transfer from hot to cold or the reverse reveals that for identical stimuli man does not respond with anywhere near the uniformity he does under conditions of constant exposure. These preliminary observations portend a new concept of human comfort in a changing environment. To understand human comfort, questions involving discomfort must be explored. There is inadequate information on man's tolerance to the extremes of cold and heat, especially the latter, when the thermal sensation may be augmented by that of pain, including man's response and tolerance to intense whole-body thermal radiation.

From the bioengineering viewpoint, a program of basic research has been initiated to develop new psychophysiological scales of comfort to better understand the pragmatic "comfort" vote used in the past with relation to environmental temperature, humidity, and air motion. Of particular interest is the role of mean radiant temperature on physiological response and comfort. To illustrate the foregoing, and without going into detail, the following is a list of titles of projects now underway:

- Theoretical Analysis of Human Temperature Regulation Using Analog Computer.
- Human Responses to Rapid Changes in Environmental Temperature.
- Temperature Regulation and Hypothalamic Neuronal Activity.
- Behavioral Responses to Heating and Cooling the Brain.
- Thermal Sweating and Vasomotor Activity.
- Electronic Instrumentation in the Acquisition of Physiological Data.
- Peripheral and Central Control of Temperature Regulation in Normal Resting Dogs.
- Temperature Regulation Studies on Human Paraplegics.
- Energy Exchange — Small Mammals.
- Body Temperature Regulation, at Onset of and During Natural Hibernation.
- Body Temperature Regulation, at Onset of and During Sleep.
- Physiological Effects of High-Intensity Radiant-Beam Heating.
- Studies on Comfort for High-Temperature Sources of Radiant Heat.
- Research in Electric Valance Heating.
- Development of an Improved Nondirectional MRT Meter.
- Physiological Bases of Thermal Discomfort.

TOTAL ENVIRONMENT

The Revolution of Urbanization

Let us leave thermal environment and go on to total environment and the questions: Where are we? Where are we going? Instead, the question could be: Where is environment taking us? When we think about man and his environment, we have to consider the broadest aspects of the problem, and include the natural environment as well as man-made environment, inside and outside of buildings.

This has become a very popular subject for discussion these days, particularly the problems that result from the rapidly accelerating urban growth. The changes in agricultural production capability, when related to the rapid expansion of our population, have brought about, in recent years, a new and potent revolution, namely, the revolution of urbanization. In this connection, I like to think of three sociotechnical revolutions that have had a tremendous impact on our environment. Many technological changes of the past 100 years have influenced our way of life, how we build our buildings, and how we communicate. But the three that I like to think about are those that have influenced the way people get about, or their mobility:

- The development of railroads for the transportation of goods and people, during the past 130 years.
- The development of the internal combustion engine, and the automobile, during the past 60 years, that provides private transportation facilities with a high degree of mobility.
- The development of mass air conditioning during the past 25 years that removed the comfort restrictions as to where people had to live or work.

There are many other technical and social developments that have influenced the way people live with their environment. However, I believe these three have had a tremendous impact on the mobility of people.

In 1960, seven out of ten Americans were urban–suburgan residents. Of our present 193 million population, more than 125 million live in 212 metropolitan areas. 84% of the growth of the 1950-1960 increase was in these same metropolitan areas. Our downtown cites have been caught between two population pressures: the rural population has tended to move to the cities, for better opportunities for jobs and education, and for conveniences, while the city people and some city industry have been moving out of the cities to the suburbs. All this as long as the railways, highways, bridges, and tunnels can provide the needed mobility.

Growth of Urban-Surburban Areas

While most of our population is now tied to the urban areas and their surrounding suburban fringes, only 1% of our land is used by them. Of course, when we add the land related to urban living, that is, land used for recreation, transportation, highways, and reservoirs, then this 1% becomes about 5%. The pressure of population increase that must be considered has recently been mentioned several times. To highlight this, we must consider that by 1990, or in the next 25 years, the situation* will be:

*Abridged Report, "Resources for the Future - 1960-2000."

- A total U.S. population predicted at 280 million, with a possible low of 250 million, and a possible high of 340 million.
- Or a predicted increase in the next 25 years of 87 millon, with a possible low increase of 57 million and a possible high increase of 147 million.
- Of this increase, the possible urban growth could be as low as 75 million and as high as 130 million, with a good possibility of another 125 million to be added in the urban–suburban metropolitan areas above the present 125 million in these urban complexes.

Obviously, with this growth, and every year we move rapidly in this direction, the total environment problems of urban life not only face us in the future, but are with us as of today, and to list but a few problems: building and buildings, housing and slum clearance, facilities for recreation, open areas near the cities, automobiles and highways, access to the cities, mass transportation, schools and hospitals, air and water pollution, sources for water, the decay of the center city and urban renewal, the urban–suburban sprawl, and, of course, the aesthetic needs for good building and good planning for the good life. And all this affects the health and well-being of 80% of our population.

Cabinet Department for Urban Affairs

President Johnson, in his budget message, called attention to this urban problem, which he called "providing a better environment," when he recommended the creation of a separate Cabinet Department to be called the Department of Housing and Urban Development. President Kennedy tried for this but was set down by the Congress. It is hoped that this new department will provide for the metropolitan areas, now so important in our national life, what the Department of Agriculture has been providing for so many years for our rural living. Senator Case discussed this question last year, and suggested that perhaps a better public response might be forthcoming if the new department were to be called the Department of Urban and Suburban Affairs, and the Department of Agriculture were renamed the Department of Agriculture and Rural Affairs.

Call it what you will, it is hoped that this new department will provide the direction and the national leadership to assist the states and local authorities in transforming a nation that for so many years was geared to a rural way of life into the now expanding complex urban–suburban way of life that involves a large percentage of our citizens as of today, and more so in the future.

The President's budget message does not provide, for the time being, that the proposed new department will encompass all activities that affect urban affairs. To illustrate, the Department of Health, Education, and Welfare will still be responsible for environmental

health and air and water pollution. This could be rightly so because
of the medical aspects of these problems. The Department of Commerce will still be responsible for the new mass transportation
(railroad) program and the building research activities of the National
Bureau of Standards.

Everyone's Problem

It would be a serious mistake to assume that the national government alone can dominate or direct the thinking as to the corrective
actions urgently needed to improve the urban–suburban situation.
Equally, it would be wrong to assume that the local governments can
handle these sprawling problems on their own.

I hope you noticed Mr. Ralph McGill's column on this subject in last
night's *Chicago Daily News* (January 28, 1965). Also, if you have not
yet read it, there is a report in this morning's *New York Times*
about a "New Center to Study Big-City Problems," with initial support
by several of the large foundations. This center would be somewhat
similar to centers established at Massachusetts Institute of Technology, Harvard, University of California, University of Chicago, and
University of Michigan, also concerned with urban problems.

These and other centers can provide the needed meeting place to
combine the forces of science and technology, industry and business,
and government and public opinion for constructive study, research,
and planning. In addition, there is a compelling need for a spreading
of the word, including programs to bring about a sense of general
involvement. Today's conference is part of this spreading of the word.
Cooperation on the part of many interested agencies is needed, such
as:

- From the private sector — by academic and research institutions,
  by architects and engineers, by business and industry, and
  by local service groups.
- From the government sector — by authorities from local, county,
  state, bistate, tristate, and quasi-governmental agencies.

There has been an apparent change in attitudes at all levels of
government for more active attention to these problems of the cities.
More interest is required at the state level because many of the
problems are beyond the directing and financial capacity of the local
governments of the cities, towns, and counties. Some change in
interest at the state capitols will result from the redistribution of
representation now under way. Increasing leadership and support by
the national government, combined with the spreading of the word,
the sense of urgency and a general involvement on the part of all,
should provide the needed action.

It has been pointed out in the course of this conference that the people and government at all levels must make up·their collective minds for corrective action now. These conferences have been occurring at rather frequent intervals, and I believe that this conference is a good step forward in highlighting the need for urgency.

Index

Contributors